GLOBAL OPTIMIZATION METHODS IN GEOPHYSICAL INVERSION

SECOND EDITION

Making inferences about systems in the earth's subsurface from remotely sensed, sparse measurements is a challenging task. Geophysical inversion aims to find models that explain geophysical observations – a model-based inversion method attempts to infer model parameters by iteratively fitting observations with theoretical predictions from trial models. Global optimization often enables the solution of non-linear models, employing a global search approach to find the absolute minimum of an objective function so that predicted data best fit the observations.

This new edition provides an up-to-date overview of the most popular global optimization methods, including a detailed description of the theoretical development underlying each method, and a thorough explanation of the design, implementation, and limitations of algorithms.

A new chapter provides details of recently developed methods, such as the neighborhood algorithm and particle swarm optimization. An expanded chapter on uncertainty estimation includes a succinct description on how to use optimization methods for model space exploration to characterize uncertainty and now discusses other new methods such as hybrid Monte Carlo and multi-chain MCMC methods. Other chapters include new examples of applications, from uncertainty in climate modeling to whole-earth studies. Several different examples of geophysical inversion, including joint inversion of disparate geophysical datasets, are provided to help readers design algorithms for their own applications.

This is an authoritative and valuable text for researchers and graduate students in geophysics, inverse theory, and exploration geoscience, and an important resource for professionals working in engineering and petroleum exploration.

MRINAL K. SEN is a Jackson Chair Professor in Applied Seismology at the University of Texas, Austin, and in January 2012, while on leave, became Director of the National Geophysical Research Institute at Hyderabad, India. His research areas include seismic wave propagation, inverse problems, seismic imaging, reservoir characterization, and computational geophysics, and he is an expert on seismic wave propagation, including anisotropy and fractures, developing analytic and numerical techniques for forward and inverse modeling. Professor Sen is the principal author of two books on geophysical inversion and co-author of over 130 papers in peer-reviewed journals and has been an instructor for several industry short courses. He serves on several national and international committees and is an associate editor of leading journals such as *Geophysics*, and *Journal of Seismic Exploration*.

PAUL L. STOFFA has held the Shell Distinguished Chair in Geophysics at the Department of Geological Sciences, University of Texas, Austin, since 1997 and was Director of the Institute for Geophysics at the same institution from 1994 to 2008. His areas of expertise are in multichannel seismic acquisition, signal processing, acoustic and elastic wave propagation, modeling, imaging, and inversion of geophysical data, and employing his knowledge of parallel computers, he works on developing new seismic data acquisition and processing methods that can be used to address complex geologic problems. Dr Stoffa has published over 100 research articles in refereed journals. A member of the Society of Exploration Geophysicists (SEG), the American Geophysical Union (AGU), and the European Association of Geoscientists and Engineers (EAGE), he is also a recipient of the Society of Brazilian Geophysicists' Foreign Geophysicist Recognition Award.

GLOBAL OPTIMIZATION METHODS IN GEOPHYSICAL INVERSION

SECOND EDITION

MRINAL K. SEN

University of Texas, Austin, USA

CSIR-National Geophysical Research Institute, India

AND

PAUL L. STOFFA

University of Texas, Austin, USA

CAMBRIDGE
UNIVERSITY PRESS

CAMBRIDGE
UNIVERSITY PRESS

University Printing House, Cambridge CB2 8BS, United Kingdom

One Liberty Plaza, 20th Floor, New York, NY 10006, USA

477 Williamstown Road, Port Melbourne, VIC 3207, Australia

4843/24, 2nd Floor, Ansari Road, Daryaganj, Delhi - 110002, India

79 Anson Road, #06-04/06, Singapore 079906

Cambridge University Press is part of the University of Cambridge.

It furthers the University's mission by disseminating knowledge in the pursuit of education, learning and research at the highest international levels of excellence.

www.cambridge.org
Information on this title: www.cambridge.org/9781108445849

First edition published by Elsevier Science B.V., 1995,
as *Global Optimization Methods in Geophysical Inversion*
First edition © Elsevier Science B.V. 1995.
Second edition © Mrinal K. Sen and Paul L. Stoffa 2013

First published by Cambridge University Press 2013
First paperback edition 2017

A catalogue record for this publication is available from the British Library

Library of Congress Cataloging in Publication data
Sen, Mrinal K.
Global optimization methods in geophysical inversion / Mrinal K. Sen, Paul L. Stoffa,
The University of Texas at Austin, Institute for Geophysics,
J.J. Pickle Research Campus. – Second edition.
pages cm
Includes bibliographical references and index.
ISBN 978-1-107-01190-8 (hardback)
1. Geological modeling. 2. Geophysics–Mathematical models. 3. Inverse problems
(Differential equations) 4. Mathematical optimization. I. Stoffa, Paul L., 1948– II. Title.
QE43.S46 2013
550.1´515357–dc23 2012033212

ISBN 978-1-107-01190-8 Hardback
ISBN 978-1-108-44584-9 Paperback

Additional resources for this publication at www.cambridge.org/sen_stoffa

Contents

Preface to the first edition (1995)

One of the major goals of geophysical inversion is to find earth models that explain geophysical observations. Thus the branch of mathematics known as *optimization* has found significant use in many geophysical applications. Geophysical inversion in this context involves finding an optimal value of a function of several variables. The function that we want to minimize (or maximize) is a misfit (or fitness) function that characterizes the differences (or similarities) between observed and synthetic data calculated by using an assumed earth model. The earth model is described by physical parameters that characterize the properties of rock layers, such as the compressional-wave velocity, shear-wave velocity, resistivity, etc.

Both *local* and *global* optimization methods are used in the estimation of material properties from geophysical data. As the title of this book suggests, our goal is to describe the application of several recently developed global optimization methods to geophysical problems. Although we emphasize the application aspects of these algorithms, we describe several parts of the theory in sufficient detail for readers to understand the underlying fundamental principles on which these algorithms are based. At this stage we take the opportunity to define some commonly used terms.

For many geophysical applications, the misfit surface as a function of the model parameters that are described by the mismatch between the predicted and observed geophysical data may be highly complicated and characterized by multiple hills and valleys. Thus such a function will have several minima and maxima; the minimum of all the minima is called the *global minimum*, and all other minima are called *local minima*. Note that the global minimum is one of the local minima, but the converse is not true, and it is also possible to have several minima of nearly the same depth. Local optimization or search algorithms such as gradient-descent methods typically attempt to find a local minimum in the close neighborhood of the starting solution. Almost all the local search methods are deterministic algorithms.

They use local properties of the misfit function to calculate an update to the current answer and search in the downhill direction. Thus these algorithms will miss the global minimum if the starting solution is nearer to one of the local minima than the global minimum. The local minimum syndrome has plagued geophysicists for over a decade now.

Recently (owing to the advent of powerful and relatively inexpensive computers), global optimization methods have been applied to several geophysical problems. Unlike local optimization methods, these methods attempt to find the global minimum of the misfit function. Most of the global optimization algorithms are stochastic in nature and use more global information about the misfit surface to update their current position. The convergence of these methods to the globally optimal solution is not guaranteed for all the algorithms. Only for some of the *simulated annealing* algorithms under certain conditions is convergence to the globally optimal solution statistically guaranteed. Also, with real observational data it is never possible to know whether the derived solution corresponds to the global minimum or not. However, our experience indicates that we are able to find many good solutions starting with only poor initial models using global optimization methods.

These global optimization methods are computationally intensive, but with the advent of vector computers, parallel computers, and powerful desktop workstations, use of these methods is becoming increasingly practical. While finding the optimal solution will always be a goal, and the global optimization methods described here are well suited for this purpose, they can also be used to obtain additional information about the nature of the solution. In particular, the description of a solution is not complete without assigning uncertainties to the derived answer. With noisy data it may not even be advisable to search for the so-called global minimum. In these situations, a statistical formulation of the inverse problem is often appealing. Consequently, we also describe how global optimization methods can be applied in a statistical framework to estimate the uncertainties in the derived result.

This is not a book on inverse theory per se; several excellent texts already exist (e.g., Menke 1984; Tarantola 1987). Our goal is to describe in sufficient detail the fundamentals of several optimization methods with application to geophysical inversion such that students, researchers, and practitioners will be able to design practical algorithms to solve their specific geophysical inversion problems. We attempted to make this book virtually self-contained so that there are no prerequisites, except for a fundamental mathematical background that includes a basic understanding of linear algebra and calculus. The material presented in this book can easily be covered in a one-semester graduate-level course on geophysical inversion. We believe that after reviewing the materials presented in this book, readers will be able to develop specific algorithms for their own applications. We will be happy to mail sample Fortran codes of prototype Metropolis simulated annealing (SA), heat bath SA, very fast simulated annealing (VFSA), and a basic genetic algorithm to those interested.

Much of our work on non-linear inversion has been supported by grants from the National Science Foundation, the Office of Naval Research, Cray Research, Inc., and the Texas Higher Education Coordinating Board. We acknowledge The University of Texas System Center for High Performance Computing for their support and computational resources. Adre Duijndam, Jacob Fokkema, Cliff Frohlich, John Goff, Lester Ingber, Tad Ulrych, and Lian-She Zhao reviewed the manuscript and offered valuable suggestions. Milton Porsani, who worked with us for one year as a visiting scientist, along with several of our graduate students, including Faruq Akbar, Carlos Calderon, Raghu Chunduru, Mike Jervis, Vik Sen, Mehmet Tanis, and Carlos Varela, participated in the research and reviewed the manuscript. Their contribution to this project has been extremely valuable. We thank Milo Backus for his many critical comments during the early stages of the work that helped tremendously in shaping our ideas on inverse theory in general. Charlene Palmer receives our thanks for painstakingly typesetting the manuscript. We also thank Gus Berkhout and Jacob Fokkema for inviting us to write the book for the series *Advances in Exploration Geophysics*.

Several figures and parts of the text in Chapter 8 are based on a paper presented at the 1994 EAEG 56th Annual Meeting and Exposition in Vienna, Austria, and have been printed by permission of the copyright holders. The copyright of the paper belongs to the European Association of Geoscientists and Engineers.

Mrinal K. Sen wishes to thank his wife, Alo, and children, Amrita and Ayon, for their sacrifice and encouragement. He also thanks Neil Frazer for his suggestion following the 1990 Society of Exploration Geophysicists Meeting to write a book on this subject.

Paul L. Stoffa wishes to thank his wife, Donna, for her constant support and Gus Berkhout, who motivated the writing of this book.

Mrinal K. Sen (mrinal@ig.utexas.edu)
Paul L. Stoffa (pauls@ig.utexas.edu)
Institute for Geophysics
The University of Texas at Austin

Preface to the second edition (2013)

The first edition of the book went out of print a few years ago. We received many requests for a copy of the book, and therefore, an invitation from Cambridge University Press to publish a revised version of this book was a welcome message. Since the publication of the first edition, a couple of review articles on Monte Carlo methods have been published in the geophysical literature. A large number of applications have also been reported. This book, however, still remains as the only publication providing a comprehensive overview of global optimization methods for geophysical applications. Global optimization methods, which were at the time of the first edition considered too slow to be of practical use, are now being used routinely in many applications. In this revised version, we have expanded several sections, including one on local optimization; we include recent global optimization methods such as the neighborhood algorithm (NA) and particle swarm optimization (PSO), and we add several new applications of global optimization methods, including the joint inversion of diverse data. Although no major new algorithmic developments of genetic algorithms (GAs) and simulated annealing (SA) have been reported, their application to solving complex problems has increased significantly since the first edition. We apologize for not being able to include all these successful applications in our reference list.

We remain indebted to our graduate students and postdoctoral fellows for collaboration. We thank our families for their unconditional support.

The University of Texas Institute for Geophysics Contribution Number 2355.

Mrinal K. Sen (mrinal@ig.utexas.edu)
Paul L. Stoffa (pauls@ig.utexas.edu)
Institute for Geophysics
The University of Texas at Austin

1

Preliminary statistics

The solution of a geophysical inverse problem can be obtained by a combination of information from observed data, the theoretical relation between data and earth parameters (models), and prior information on data and models. Due to uncertainties in the data and model, probability theory can be used as a tool to describe the inverse problem. Excellent introductory books on the subject of probability theory are those of Feller (1968), Papoulis (1965), and Ross (1989). In this chapter we review probability theory and stochastic processes, concepts that are used later to describe the global optimization methods used in geophysical inverse problems. Readers familiar with the subject can proceed directly to Chapter 2.

1.1 Random variables

In simple language, a random variable is a variable used to represent the outcome of a random experiment. Familiar random experiments are the tossing of a die and the flipping of a coin. When a die is tossed, there are six possible outcomes, and it is not certain which one will occur. Similarly, when a coin is tossed, there are two possible outcomes, and it is not certain which one will occur. The outcome of a random experiment is usually represented by a point called a *sample point s*. The set that consists of all possible sample points is called the *sample space S*. Subsets of *S* represent certain events such that an event *A* consists of a certain collection of possible outcomes *s*. If two subsets contain no point *s* in common, they are said to be *disjoint*, and the corresponding events are said to be *mutually exclusive*. Formally, any single-valued numerical function $X(s)$ defined on a sample space *S* is called a *random variable*, and a unique real number is associated with each point *s*.

1.2 Random numbers

Most of the methods of geophysical inversion that we discuss in this book use probability or statistical theories that involve studying processes arising from random experiments. This means that we need to simulate random processes on a computer. In practice, this requires an algorithm to generate random numbers. Computer-generated random numbers have been used extensively in several applications. Most commonly, sampling methods using computer-generated random numbers have been used in situations where the mathematics became intractable. Metropolis and Ulam (1949) first proposed the application of random sampling to the solution of deterministic problems such as the evaluation of an integral of the type

$$I = \int_{x_{\min}}^{x_{\max}} f(x) \, dx. \tag{1.1}$$

If this integral exists, it is given by the expected value of the function. For a random number X that has a uniform distribution over the interval $(x_{\min}, x_{\max})$, the preceding integral can be replaced by the following sum:

$$I = \frac{1}{n} \sum_{i=1}^{i=n} f(X_1), \tag{1.2}$$

where X_i is a random sample from the uniform distribution. Techniques using this idea are referred to as *Monte Carlo methods*. In general, any method using a random walk is usually included in the category of Monte Carlo methods.

To apply any technique that involves random processes, computer-generated random numbers are required. Although a computer is a machine that produces output that is always predictable, computer-generated random numbers are used extensively and are called *pseudorandom numbers*. Most computers use a method called the *congruential method* to generate random samples from a uniform distribution (Kennedy and Gentle 1980; Rubinstein 1981). The machine produces a sequence of pseudorandom integers $I_1, I_2, I_3, \ldots$ between 0 and N by the recurrence relation

$$I_{j+1} = \mathrm{mod}(aI_j + c, N), \tag{1.3}$$

where the mod operation is defined as

$$\mathrm{mod}(a_1, a_2) = a_1 - \mathrm{int}\left(\frac{a_1}{a_2}\right) a_2. \tag{1.4}$$

In Eq. (1.3), a and c are positive integers called the *multiplier* and *increment*, respectively. The number thus generated will repeat itself with a period no greater than N. If N, a, and c are properly chosen, then the period will be of maximum length; i.e., all possible integers between 0 and N will occur at some point. The highest possible value of N is controlled by the precision of the machine used. The real number that is returned is given by (I_{j+1}/N), so the number lies between 0 and 1. The random numbers thus generated are not free from sequential correlation in successive calls. Also, improper choice of the parameters N, a, and c causes further problems. Press *et al.* (1989) describe methods to improve the performance of random number generators. One of them is to do additional randomizing shuffles on the numbers generated by the machine. Given random samples from the standard uniform distribution, random numbers from some other distribution may be readily obtained by transformations (Rubinstein 1981).

1.3 Probability

The concept of probability is used to analyze the experiment represented by set S. Let us assume that $P\{A\}$ denotes the probability of an event (collection of outcomes) A; then we make the following axioms:

$$P\{A\} \geq 0 \quad \text{for any event } A; \tag{1.5}$$

i.e., the probabilities of all the events are non-negative and

$$P\{S\} = 1; \tag{1.6}$$

i.e., the probability of the whole sample space is unity.

Thus, by convention or axiom, the probability that one of a sequence of mutually exclusive events $\{Ai\}$ will occur is given by

$$P\{A_1 \text{ or } A_2 \text{ or}...\} = \sum_i P(A_i). \tag{1.7}$$

There exist different interpretations of the concept of probability. They can be classified into the following categories: (1) the classical interpretation, (2) the relative-frequency interpretation, and (3) the Bayesian interpretation. The classical definition of probability (Papoulis 1965) is as follows: Let N be the total number of possible outcomes of an experiment. If in N_A of all these outcomes the event A occurs, then $P(A)$ is given by

$$P(A) = \frac{N_A}{N}, \tag{1.8a}$$

provided that the occurrence of all the events is equally likely. The main criticism of this definition is that it is circular because *equally likely* also means *equally probable*.

According to the relative-frequency interpretation (von Mises 1957), the probability of an event A is the following limit of the relative frequency:

$$P(A) = \lim_{N \to \infty} \frac{N_A}{N}, \qquad (1.8b)$$

where N_A is the number of occurrences of A in N number of trials. In the relative-frequency definition, the concept of equally likely events is completely avoided. Here a trial is used to mean repetition of an experiment under identical circumstances. The problems with the relative-frequency interpretations are as follows: The limit (Eq. 1.8b) can only be assumed to exist, the number of trials is always finite, and the definition gives no meaning to the probability of a hypothesis (Jeffreys 1939).

The Bayesian interpretation of probability is due to Bayes (1763), who defined the probability to be a *degree of belief*. Thus the probability theory can be viewed as an extension of deductive logic and is called *inductive logic*. In *deductive logic*, a proposition can either be true or false, but in *inductive logic*, the probability of a proposition constitutes a degree of belief, with proof or disproof as extremes.

The Bayesian interpretation can again be classified into two categories: the logical interpretation and the subjective interpretation. In the logical interpretation, the probability is objective, an aspect of the *state of affairs*. In the subjective interpretation, the degree of belief is a personal degree of belief such that the axioms of probability theory are not violated.

Most statisticians are proponents of one interpretation or another, but more than one interpretation may be helpful because different situations simply may ask for different interpretations. Leaving aside the question of whether the logical or subjective interpretation is preferable, a Bayesian interpretation of probability appears to be conceptually very clear.

1.4 Probability distribution, distribution function, and density function

A discrete random variable $X(s)$ assumes a finite number of values. For each value x_i, there is a unique probability that the random variable assumes the value x_i:

$$P\{X(s) = x_i\} = p_i, \quad i = 0, 1, 2, \dots . \qquad (1.9)$$

The sequence $\{p_i\}$ is called the *probability distribution* of $X(s)$, and the cumulative probability

$$P\{X(s) \le x\} = \sum_{x_i \le x} p_i = F(x), \quad -\infty < x < \infty, \qquad (1.10)$$

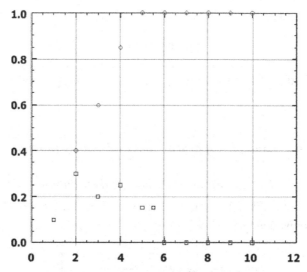

Figure 1.1 An example of a probability distribution (*open squares*) and its corresponding distribution function (*open circles*). Color version available online at www.cambridge.org/sen_stoffa.

is called the *distribution function* of $X(s)$. For example, consider that for a random variable assuming values x_i, $i = 0, 1, 2, 3, 4, 5, \ldots$, the probability distribution is given as $p_0 = 0$, $p_1 = 0.1$, $p_2 = 0.3$, $p_3 = 0.2$, $p_4 = 0.25$, and $p_5 = 0.15$ and 0 elsewhere. The distribution function $F(x)$ can be computed using Eq. (1.10). The probability distribution and corresponding distribution function are shown in Figure 1.1.

The following properties are satisfied by a distribution function: $F(-\infty) = 0$, $F(\infty) = 1$, and it is a non-decreasing function of x: $F(x_1) = F(x_2)$ for $x_1 < x_2$. If $F(x)$ is differentiable, the probability density function (pdf) $p(x)$ is given by

$$p(x) = \frac{dF(x)}{dx}.$$ (1.11a)

If $F(\mathbf{m})$ denotes the distribution function of a vector of variables $\mathbf{m}$, with $\mathbf{m} = [m_1, m_2, \ldots, m_n]^T$, then if $F(\mathbf{m})$ is differentiable, the probability density function (pdf) $p(\mathbf{m})$ of $\mathbf{m}$ is given by

$$p(\mathbf{m}) = \frac{\partial^n F(\mathbf{m})}{\partial m_1 \, \partial m_2 \cdots \partial m_n}.$$ (1.11b)

The probability of $\mathbf{m}$ being in a certain region or volume A is given by

$$P(\mathbf{m} \in A) = \int_A p(\mathbf{m}) \partial m_1 \, \partial m_2 \cdots \partial m_n.$$ (1.12)

The strict definition of the pdf is that it is normalized; i.e.,

$$\int p(\mathbf{m})\, d\mathbf{m} = 1. \tag{1.13}$$

An unnormalized probability density function is simply called the *density function*.

1.4.1 Examples of distribution and density functions

1.4.1.1 Normal or Gaussian distribution

A continuous random variable x is said to be *normally distributed* if its density function $p(x)$ is given as

$$p(x) = \frac{1}{\sigma\sqrt{2\pi}} \exp\left[-\frac{(x - <x>)^2}{2\sigma^2}\right], \tag{1.14}$$

where $<x>$ and σ^2 are called the *mean* and the *variance*, respectively. Note that

$$\int_{-\infty}^{\infty} dx\, p(x) = 1. \tag{1.15}$$

The corresponding distribution function is given by

$$F(x) = \int_{-\infty}^{x} dx' p(x') = \frac{1}{2} + \mathrm{erf}\left(\frac{x - <x>}{\sigma}\right), \tag{1.16}$$

where erf(x) is called the *error function* (Abramowitz and Stegun 1972). The plots of the Gaussian probability density and distribution functions are shown in Figure 1.2. Note that the mean and the variance completely describe a normal distribution.

1.4.1.2 Cauchy distribution

The *Cauchy density function* for a continuous random variable x is defined as

$$p(x) = \frac{\alpha/\pi}{\alpha^2 + x^2}, \tag{1.17}$$

where α is a constant that controls the width of the distribution (Figure 1.3). The peak value of the distribution is equal to $(1/\pi\alpha)$; i.e., it is inversely proportional to the value of α. Thus, with decreasing values of α, the distribution becomes narrower, and the height of the peak increases.

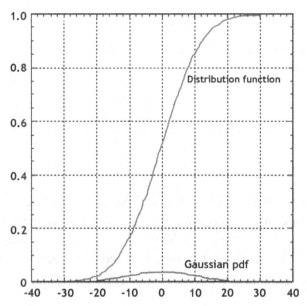

Figure 1.2 A Gaussian probability density function for a mean = 0.0 and a standard deviation = 10.0. Its corresponding distribution function is also shown. Color version available online at www.cambridge.org/sen_stoffa.

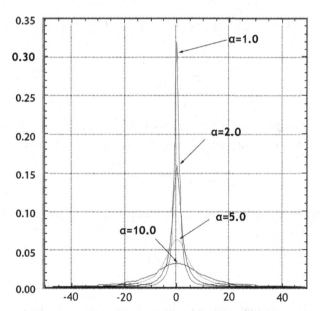

Figure 1.3 A Cauchy probability density function for different values of α. Color version available online at www.cambridge.org/sen_stoffa.

1.4.1.3 Gibbs' distribution

For a continuous random variable x, the *Gibbs probability density function* (also called the *Boltzmann pdf*) is defined as

$$p(x) = \frac{\exp\left[\dfrac{E(x)}{T}\right]}{\displaystyle\int_{-\infty}^{\infty} dx' \exp\left[\dfrac{E(x')}{T}\right]}, \qquad (1.18)$$

where $E(x)$ is a function of the random variable x, and T is a control parameter. Note that the shape of the distribution is controlled by changing the parameter T. This parameter is called *temperature* in thermodynamics. This density function is not analytically integrable except in some very simple cases. This density function will be discussed in much greater detail in subsequent chapters.

1.5 Joint and marginal probability distributions

Let X and Y be two discrete random variables defined on the same sample space. Their joint (bivariate) probability distribution $\{\Gamma_{ij}\}$ is defined by

$$\Gamma_{ij} = P\{X = x_i \text{ and } Y = y_j\}, \qquad i, j = 0, 1, \ldots \qquad (1.19)$$

with

$$\sum_i \sum_j \Gamma_{ij} = 1. \qquad (1.20)$$

The sequence $\{p_i\}$ and $\{q_j\}$, called the *marginal distributions* of X and Y, respectively, are calculated from the joint distribution by the following formulas:

$$p_i = \sum_j \Gamma_{ij} \quad \text{and} \quad q_j = \sum_i \Gamma_{ij}. \qquad (1.21)$$

When two random variables X and Y are independently distributed, or stochastically independent, we have

$$\Gamma_{ij} = p_i q_j, \qquad (1.22)$$

for all i and j.

Definitions for marginal distributions of continuous variables can be written down simply by replacing the sums in Eq. (1.21) with integrals. Also, equations similar to Eq. (1.21) can be easily written down for joint and marginal distribution and density functions.

1.6 Mathematical expectation, moments, variances, and covariances

The expectation, moments, variances, and covariances are the quantities usually required to describe a distribution. The *mathematical expectation* (or *mean value*, or *expected value*) of a discrete random variable X is defined as

$$e(X) = <x> = \sum_i x_i p_i \tag{1.23}$$

and of a continuous random variable as

$$e(X) = <x> = \int dx \, x p(x). \tag{1.24}$$

The mean $e(X)$ exists if the sum or the integral converges absolutely. The mean or the mathematical expectation is also called the *first moment*. A moment of order r is given by $e[Xr]$. The central moment of order r is given by $e\{[X - e(X)]r\}$, $r = 1$, $2,\dots$. That is, the second central moment (also called the *variance* of the distribution) is given by

$$\sigma_x^2 = e\left\{[X - e(X)]^2\right\} = \int_{-\infty}^{\infty} dx (x - <x>)^2 p(x). \tag{1.25}$$

The square root of the variance is called the *standard deviation*.

Let us now consider two continuous random variables X and Y having a joint distribution such that the moments of the marginal distributions describe the properties of X and Y, respectively. X and Y are now related using the covariance of X and Y as given by

$$\begin{aligned}\sigma_{X,Y} &= e\left\{[X - e(X)] - [Y - e(Y)]\right\} \\ &= e(XY) - e(X)E(Y).\end{aligned} \tag{1.26}$$

Now consider two random variables X and Y with standard deviations σ_X and σ_Y, respectively, and define the following:

$$U = \frac{X - e(X)}{\sigma_X} \quad \text{and} \quad V = \frac{Y - e(Y)}{\sigma_Y}; \tag{1.27}$$

i.e., $e(U) = 0$ and $e(V) = 0$ and $\sigma_u^2 = e(U^2) = 1$ and $\sigma_v^2 = e(V^2) = 1$. U and V are called *standardized random variables*. The *correlation coefficient* between X and Y is defined as the covariance between U and V; i.e.,

$$\rho_{XY} = \sigma_{U,V} = e(UV) = E\left\{\left[\frac{X - e(X)}{\sigma_X}\right]\left[\frac{Y - e(Y)}{\sigma_Y}\right]\right\}.$$

Using Eq. (1.26), we obtain

$$\rho_{XY} = \frac{\sigma_{X,Y}}{\sigma_X \sigma_Y}. \tag{1.28}$$

The value of the correlation coefficient is not affected by changes of scale or origin because it is calculated from standardized random variables.

The covariance function is a measure of correlation between the values of the random variables; i.e., it determines if the values taken by one random variable depend on the values of another. Consider a vector

$$\mathbf{x} = [x_1, x_2, x_3, \ldots, x_n]^T, \tag{1.29}$$

of n random variables x_1, x_2, x_3, ..., x_n with means $<x_1>$, $<x_2>$, $<x_3>$, ..., $<x_n>$, respectively. Assuming that these random variables are continuous, the mean values are given by the following integral:

$$<\mathbf{x}> = \int_{-\infty}^{\infty} d\mathbf{x}\, \mathbf{x}\, p(\mathbf{x}), \tag{1.30}$$

or, for each variable, as

$$\langle x_i \rangle = \int_{-\infty}^{\infty} \partial x_1 \int_{-\infty}^{\infty} \partial x_2 \cdots \int_{-\infty}^{\infty} \partial x_n\, x_i p(\mathbf{x}). \tag{1.31}$$

Similarly, the covariance matrix is given by

$$[\mathrm{cov}(\mathbf{x})] = \int_{-\infty}^{\infty} d\mathbf{x}(\mathbf{x} - <\mathbf{x}>)(\mathbf{x} - <\mathbf{x}>)^T p(\mathbf{x}), \tag{1.32}$$

or each component will be given by the following equation:

$$[\mathrm{cov}(\mathbf{x})]_{ij} = \int_{-\infty}^{\infty} dx_1 \int_{-\infty}^{\infty} dx_2 \cdots \int_{-\infty}^{\infty} dx_n (x_i - \langle x_i \rangle)(x_j - \langle x_j \rangle) p(\mathbf{x}). \tag{1.33}$$

The joint distribution of two independent random variables is simply the product of two univariate distributions. When the variables are correlated (characterized by the covariance matrix), the distribution is more complicated, and a general form of the normal distribution for a vector of random variable $\mathbf{x}$ is defined as (Menke 1984)

$$p(\mathbf{x}) = \frac{|\mathrm{cov}\ \mathbf{x}|^{-1/2}}{(2\pi)^{n/2}} \exp\left\{ -\frac{1}{2}[\mathbf{x} - <\mathbf{x}>]^T [\mathrm{cov}\ \mathbf{x}]^{-1} [\mathbf{x} - <\mathbf{x}>] \right\}. \tag{1.34}$$

1.7 Conditional probability and Bayes' rule

Let us assume that the events A_1 and A_2 are among the possible outcomes of an experiment and that $P\{A_1\}$ and $P\{A_2\}$ are the probabilities of the two events, respectively. According to the classical definition of probability,

$$P\{A_1\} = \frac{\text{sum of probabilities of sample points in } A_1}{\text{sum of probabilities of all points in the sample space}}. \tag{1.35}$$

We also use the symbol $P\{A_1, A_2\}$ to represent joint probability of events A_1 and A_2, as defined in Eq. (1.19).

Next, we assume that if A_2 has occurred, then the symbol $P\{A_1|A_2\}$ stands for the *conditional probability* of event A_1 given that A_2 has occurred. In such a situation, the outcome of the experiment must be in a more restricted region than the entire sample space. Based on the axioms of probability, we can write

$$P\{A_1|A_2\} = \frac{P\{A_1, A_2\}}{P\{A_2\}}, \quad \text{assuming that } P\{A_2\} \neq 0. \tag{1.36}$$

Thus

$$P\{A_1, A_2\} = P\{A_2\} P\{A_1|A_2\}. \tag{1.37}$$

Again assuming that $P\{A_1|A_2\}$ represents the conditional probability of event A_2 given that A_1 has occurred, we can write

$$P\{A_2|A_1\} = \frac{P\{A_2, A_1\}}{P\{A\}}, \quad \text{assuming that } P\{A_1\} \neq 0. \tag{1.38}$$

The conditional probability distribution and density functions can now easily be defined. The conditional distribution function $F(x|x_0)$ of a random variable $X(s)$, assuming x_0, is defined as the conditional probability of the event $\{X(s) = x|x_0\}$; i.e.,

$$F(x|x_0) = P\{X(s) \leq x|x_0\} = \frac{P\{X(s) \leq x, x_0\}}{P\{X_0\}}, \tag{1.39}$$

where $\{X(s) \leq x, x_0\}$ is the event consisting of all outcomes ε such that $X(\varepsilon) \leq x$ and $\varepsilon \in x_0$

If we consider two vectors of random variables $\mathbf{x}$ and $\mathbf{y}$ such that their joint probability density function is denoted by $p(\mathbf{x}, \mathbf{y})$, the marginal pdf of $\mathbf{x}$ will be given as

$$p(\mathbf{x}) = \int d\mathbf{y}\, p(\mathbf{x}, \mathbf{y}), \tag{1.40}$$

which can be regarded as the pdf of **x** ignoring or averaging over variable **y**. The conditional pdf $p(\mathbf{x}|\mathbf{y})$ called the *pdf of* **x** *for given values of* **y** is defined as

$$p(\mathbf{x}|\mathbf{y}) = \frac{p(\mathbf{x}, \mathbf{y})}{p(\mathbf{y})}. \tag{1.41}$$

Thus we have

$$p(\mathbf{x}, \mathbf{y}) = p(\mathbf{x}|\mathbf{y}) p(\mathbf{y}). \tag{1.42}$$

Also,

$$p(\mathbf{y}|\mathbf{x}) = \frac{p(\mathbf{x}, \mathbf{y})}{p(\mathbf{x})}, \tag{1.43}$$

or

$$p(\mathbf{x}, \mathbf{y}) = p(\mathbf{y}|\mathbf{x}) p(\mathbf{x}). \tag{1.44}$$

From Eqs. (1.42) and (1.44), we have

$$p(\mathbf{x}|\mathbf{y}) p(\mathbf{y}) = p(\mathbf{y}|\mathbf{x}) p(\mathbf{x}), \tag{1.45}$$

which gives

$$p\left[\mathbf{x}|\mathbf{y} = \frac{p(\mathbf{y}|\mathbf{x}) p(\mathbf{x})}{p(\mathbf{y})}\right]. \tag{1.46}$$

The preceding relation is called *Bayes' rule* in statistics and is widely used in problems of model parameter estimation by data fitting. If **x** is identified with the model and **y** with the data vector, $p(\mathbf{x}|\mathbf{y})$, the probability of **x** given measurements of **y**, is expressed as a function of $p(\mathbf{x})$, the marginal pdf of the model independent of data and $p(\mathbf{y}|\mathbf{x})$, the conditional probability of **y** given **x**. $p(\mathbf{x}|\mathbf{y})$ is called the *posterior* or *a posteriori pdf*, whereas $p(\mathbf{x})$ is called the *prior* or *a priori pdf* and contains the prior information about the model independent of measurements. The marginal density function $p(\mathbf{y})$ is usually assumed to be a constant, and the conditional distribution $p(\mathbf{y}|\mathbf{x})$ is usually called a *likelihood function*. The interpretation of Bayes' theorem is as follows: The state of information on **x** and **y** is described by the joint pdf $p(\mathbf{x},\mathbf{y})$. Information becomes available as values of **y** are obtained. So the question we ask is, "How should the pdf of **x** be calculated in this situation?" According to the definitions of conditional probability, this pdf should be proportional to $p(\mathbf{y}|\mathbf{x})$ with the obtained values for **y** substituted. The final expression for the conditional pdf $p(\mathbf{x}|\mathbf{y})$ thus is given by Eq. (1.46). Bayes' rule is especially appealing because it provides a mathematical formulation of how current knowledge can be updated when new information becomes available.

1.8 Monte Carlo integration

After the preceding discussions of random numbers, random variables, and pdfs, we revisit the problem of evaluating the integral in Eq. (1.1) by Monte Carlo methods. The simplest of these methods is known as the *hit-or-miss Monte Carlo method* (Rubinstein 1981). Recall the integral (Eq. 1.1)

$$I = \int_{x_{min}}^{x_{max}} f(x)\,dx,$$

and let us assume that $y = f(x)$ is bounded such that $0 \le f(x) \le y_{max}$ and $x_{min} \le x \le x_{max}$. We also define a random vector (X, Y) with a uniform pdf; i.e.,

$$p_{x,y}(x,y) = \begin{cases} \dfrac{1}{y_{max}(x_{max} - x_{min})}, & \text{if } x_{min} \le x \le x_{max} \text{ and } 0 \le y \le y_{max}, \\ 0 & \text{elsewhere.} \end{cases} \quad (1.47)$$

The probability that a (X, Y) will fall within the area defined by Eq. (1.1) is given by

$$p = \frac{I}{y_{max}(x_{max} - x_{min})}. \quad (1.48)$$

If in N trials from the distribution p_{xy}, N_H is the number of trials within the area bounded by the curve $f(x)$, the approximate value of p will be given as

$$\hat{p} = \frac{N_H}{N}, \quad (1.49)$$

Thus we have

$$I \approx y_{max}(x_{max} - x_{min})\frac{N_H}{N}. \quad (1.50)$$

Thus, by taking a large sample N and counting the number of hits N_H, we can evaluate I. It can be shown (Rubinstein 1981) that the precision of the estimator of the integral as given by the preceding equation measured by its standard deviation is of the order $N^{-1/2}$. Thus a large size N would result in a good estimate of the integral. Rubinstein (1981, p. 117) also showed how to choose N for a desired precision. Note that the hit-or-miss Monte Carlo method is commonly known as the *Monte Carlo method*.

1.9 Importance sampling

The hit-or-miss Monte Carlo method draws samples from a uniform distribution (e.g., Eq. 1.47). If the integrand only takes on significant values in some small regions, many of these samples do not significantly contribute to the integral. The idea of importance sampling is to concentrate the distribution of sample points in regions that are most important, i.e., that contribute the most to the integral. This requires that unlike the hit-or-miss Monte Carlo method, the random vector must be drawn from a non-uniform distribution. Following Rubinstein (1981), we will examine what pdf should be used in lieu of Eq. (1.47) such that the variance of the estimator of the integral is minimum.

Let us rewrite Eq. (1.1) in the following form:

$$I = \int \frac{f(x)}{p_x(x)} p_x(x)\, dx, \tag{1.51}$$

where the random value x is characterized by the pdf $p_x(x)$ such that $p_x(x) > 0$ for all x in the region. Clearly, following Eq. (1.24), the preceding equation gives the expected value of $\zeta = f(x)/p_x(x)$; i.e.,

$$I = E(\zeta) = E\left[\frac{f(x)}{p_x(x)}\right]. \tag{1.52}$$

The function $p_x(x)$ is called the *importance sampling distribution*. The variance of the estimator ζ is given by

$$\mathrm{var}(\zeta) = \int \left[\frac{f(x)}{p_x(x)} - I\right]^2 p_x(x)\, dx,$$

or

$$\mathrm{var}(\zeta) = \int \frac{[f(x)]^2}{p_x(x)}\, dx - I^2. \tag{1.53}$$

Also, we know that we can estimate the integral by taking samples $x_1, \ldots, x_N$ from pdf $p_x(x)$ and using the following relation for the expected value of ζ

$$\theta = \frac{1}{N} \sum_{i=1}^{N} \frac{f(x_i)}{p_x(x_i)}, \tag{1.54}$$

when θ is an estimate of $E(\zeta)$.

We now need to know what the pdf $p_x(x)$ must look like in order to have a minimum of variance ζ. It can be shown by using the Cauchy–Schwartz inequality that the minimum of variance of ζ occurs at

$$\mathrm{var}(\zeta) = \left[\int |f(x)| dx\right]^2 - I^2. \tag{1.55}$$

This occurs for

$$p_x(x) = \frac{|f(x)|}{\int |f(x)| dx}. \tag{1.56}$$

We can show that the minimum of variance (Eq. 1.55) can be obtained by substituting $p_x(x)$ in the equation for var(ζ). If $f(x) > 0$, then the pdf that is optimal for the evaluation of this integral is

$$p_x(x) = \frac{f(x)}{I}, \tag{1.57}$$

and var(ζ) = 0. This means that in order to evaluate the integral with minimum variance, the random vectors must be drawn from a distribution that is proportional to the function $f(x)$ itself. But does this solve the problem? Not really! Note that $p_x(x)$ now contains the integral I in the denominator. And if we know the value of I, there is no need for us to use any importance sampling technique to evaluate I.

The simulated annealing (SA) method described in detail in Chapter 4 exactly addresses this issue by using the concept of a Markov chain to attain a stationary distribution asymptotically, and thus it can be used as an importance sampling technique. Importance sampling is more easily implemented with Markov chain methods because it is not necessary to construct the distribution $p_x(x)$. Consequently, in the following sections we describe the stochastic processes, Markov chains, and other related concepts.

1.10 Stochastic processes

Consider an experiment with its outcome ζ forming the space S. We now assign to every value of ζ an index t and define the function

$$X(t, \zeta),$$

where the index t is often interpreted as time. We now have a family of functions for different times t, one for each ζ. This family is called a *stochastic process*. It can be viewed as a function of two variables. For a specific ζ_i, it represents a

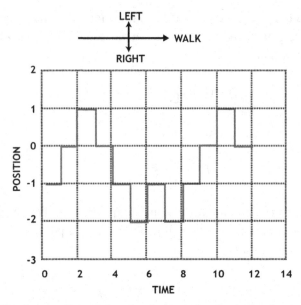

Figure 1.4 Random-walk model: a classic example of a stochastic process. Color version available online at www.cambridge.org/sen_stoffa.

single time function, whereas for a given time t_i, it represents a random variable. Dropping the ζ term from our notation, a stochastic process

$$\{X(t), t \in T\}$$

is referred to as the *state* of the process at time t. Several examples of stochastic processes can be found in Papoulis (1965) and Kemeny and Snell (1960). The set T is called the *index set* of the process. For discrete T, i.e., when T is a countable set, the stochastic process is called a *discrete-time process*; otherwise, it is called a *continuous-time process*. The space containing all possible values of the random variables $X(t)$ is called the *state space*.

   Evolution of some physical process through time can be described by means of a stochastic process. A classic example of a stochastic process is a random walk (e.g., Papoulis 1965), in which an unbiased coin is tossed and a person takes a step to the left if a head (H) occurs or to the right if a tail (T) occurs. The person's position t seconds after the tossing started will be denoted by $X(t)$, which clearly depends on the particular sequences of heads and tails. For unit time step with outcomes *THHTTTHTHHHT*, the random walk $X(t)$ can be drawn as shown in Figure 1.4.

   For a specific t, $X(t)$ is a random variable. Therefore, the definitions of the probability density and distribution functions discussed earlier will also apply.

However, they depend on the index t. For example, the distribution function will be given as

$$F(x;t) = P\{X(t) \le x\}. \tag{1.58}$$

When we consider two random variables $X(t_1)$ and $X(t_2)$ at two times t_1 and t_2, respectively, their joint distribution function will obviously depend on both t_1 and t_2 and is given by

$$F(x_1, x_2; t_1, t_2) = P\{X(t_1) \le x_1; X(t_2) \le x_2\}. \tag{1.59}$$

Similarly, following the earlier conventions, $p(x_1, x_2; t_1, t_2)$ represents the joint probability density function. The conditional probability density $p(x_1; t_1 | X(t_2) = x_2)$ now represents the probability that the random variable at time t_1 has a value x_1 given that its value is x_2 at time t_2 and is given by

$$p(x_1; t_1 | X(t_2) = x_2) = \frac{p(x_1, x_2; t_1, t_2)}{p(x_2; t_2)}. \tag{1.60}$$

1.11 Markov chains

Let us consider a stochastic process given by $X(k)$ that represents the outcome of the kth trial or time step. We also assume that it can take a finite number of possible values. If $X(t) = i$, we will say that the process is in state i at time t. We also define a probability $P_{ij}(t + 1)$ to represent the probability that a stochastic process will be in state j at time $t + 1$ given that it is in state i at time t (recall that each possible outcome is called a *state*); i.e.,

$$P_{ij}(t+1) = P\{X(t+1) = j | X(t) = i\}. \tag{1.61}$$

A *Markov chain* is a stochastic process in which the conditional distribution at any future time $t + 1$ for the given past states and the present state is independent of the past states and depends only on the present state; i.e.,

$$\begin{aligned} P_{ij}(t+1) &= P\{X(t+1) = j | X(t) = i, X(t-1) = i_{t-1}, \ldots, X(0) = i_0\} \\ &= P\{X(t+1) = j | X(t) = i\}, \end{aligned} \tag{1.62}$$

Thus the probability of the outcome of a given trial depends only on the outcome of the previous trial and not on any other. The quantity P_{ij} is called the *transition probability* and is one element of the transition probability matrix **P** or, more precisely, an element of a single-step transition probability matrix given by

$$\mathbf{P} = \begin{bmatrix} P_{00} & P_{01} & P_{02} & \cdot & \cdot \\ P_{10} & P_{11} & P_{12} & \cdot & \cdot \\ \cdot & \cdot & \cdot & \cdot & \cdot \\ \cdot & \cdot & \cdot & \cdot & \cdot \\ P_{i0} & P_{i1} & P_{i2} & \cdot & \cdot & \cdot \\ \cdot & \cdot & \cdot & \cdot & \cdot & \cdot \end{bmatrix},$$

where

$$P_{ij} \geq 0, \quad \text{for all } i, j \text{ such that } i, j \geq 0, \tag{1.63}$$

and

$$\sum_{j=0}^{\infty} P_{ij} = 1, \quad i = 0, 1, 2, \ldots. \tag{1.64}$$

A matrix with these properties is called a *stochastic matrix*.

As an example, let us consider the case where we have four possible states, and $P_{12}(k)$ represents a single-step probability of transition from state 1 to state 2 at step k. Therefore,

$$\mathbf{P}(k) = \begin{bmatrix} P_{11} & P_{12} & P_{13} & P_{14} \\ P_{21} & P_{22} & P_{23} & P_{24} \\ P_{31} & P_{32} & P_{33} & P_{34} \\ P_{41} & P_{42} & P_{43} & P_{44} \end{bmatrix}. \tag{1.65}$$

Naturally,

$$\begin{aligned} P_{11} + P_{12} + P_{13} + P_{14} &= 1.0, \\ P_{21} + P_{22} + P_{23} + P_{24} &= 1.0, \\ P_{31} + P_{32} + P_{33} + P_{34} &= 1.0, \\ P_{41} + P_{42} + P_{43} + P_{44} &= 1.0. \end{aligned} \tag{1.66}$$

Let $a_i(k)$ denote the probability of outcome i at the k trial; i.e., $a_i(k)$ is the probability that a state i will occur at time step k. Thus $a_i(k)$ is the element of a state probability vector

$$\mathbf{a}(k) = \begin{bmatrix} a_1 \\ a_2 \\ a_3 \\ a_4 \end{bmatrix}.$$ (1.67)

The state probability vector at any time step k is given by the product of the state probability vector at time step $k - 1$ and the transition probability matrix $\mathbf{P}(k)$; i.e.,

$$a_{i(k)} = \sum_l a_l (k-1) P_{li}(k).$$ (1.68)

For example,

$$\left[a_1(k) = a_1 (k-1) P_{11}(k) + a_2 (k-1) P_{21}(k) + a_3 (k-1) P_{31}(k) + a_4 (k-1) P_{41}(k) \right],$$

or, in matrix notation, we have

$$\mathbf{a}(k) = \mathbf{a}(k-1)\mathbf{P}(k).$$ (1.69)

Thus, if $\mathbf{a}(0)$ represents the initial state probability vector, then

$$\begin{aligned} \mathbf{a}(1) &= \mathbf{a}(0)\mathbf{P}(1) \\ \mathbf{a}(2) &= \mathbf{a}(1)\mathbf{P}(2) = \mathbf{a}(0)\mathbf{P}(1)\mathbf{P}(2) \\ \mathbf{a}(3) &= \mathbf{a}(2)\mathbf{P}(3) = \mathbf{a}(0)\mathbf{P}(1)\mathbf{P}(2)\mathbf{P}(3) \\ &\vdots \\ \mathbf{a}(k) &= \mathbf{a}(0)\mathbf{P}(1)\mathbf{P}(2)\mathbf{P}(3)...\mathbf{P}(k). \end{aligned}$$ (1.70)

Thus, given the state probability vector at the beginning of the process, the probability of the outcome at any step can be computed using the transition probability matrices. When the transition probability matrices do not change with time, i.e.,

$$\mathbf{P} = \mathbf{P}(1) = \mathbf{P}(2) = \mathbf{P}(3) = \cdots = \mathbf{P}(k),$$ (1.71)

we have

$$\mathbf{a}(k) = \mathbf{a}(0)(\mathbf{P})^k.$$ (1.72)

Earlier we defined a one-step transition probability P_{ij}. Similarly, an n-step transition probability P_{ij}^n is the probability that the process that is in state i will be in state j after n additional steps; i.e.,

$$P_{ij}^n = P\{X_{n+m} = j | x_m = i\}, \quad n \geq 0.$$ (1.73)

The Chapman–Kolmogorov equation (Ross 1989)

$$P_{ij}^{n+m} = \sum_{k=0}^{\infty} P_{ik}^{n} P_{kj}^{m} \ \forall \ n,m > 0, \tag{1.74}$$

can be used to calculate the $n + m$ step transition probabilities in terms of an n-step transition probability P_{ik}^{n} and an m-step transition probability P_{kj}^{m}.

1.12 Homogeneous, inhomogeneous, irreducible, and aperiodic Markov chains

If the transition probabilities $P_{ij}(k)$ do not depend on the time step k – i.e., they do not change with time – the Markov chain is said to be *homogeneous*; otherwise, it is called *inhomogeneous*. Thus a homogeneous Markov chain is also a stationary Markov chain.

The states of a Markov chain can be classified based on whether the process can go from a state i to a state j (not necessarily in one step). A state j is said to be accessible from state i if $P_{ij}^{n} > 0$ for some $n = 0$. This means that there is a finite probability that the process will be in a state j, starting from a state i. Two states i and j that are accessible to each other are said to *communicate*, which means that we can go from either state to the other one. Such states are said to belong to the same equivalence class (Kemeny and Snell 1960).

A Markov chain is said to be *irreducible* if there is only one equivalence class; i.e., all the states can communicate with each other. A set of states in which all members of the set are reachable (over time and with positive probability) from all other members of the set is called an *ergodic class*.

For any state i, let us denote P_{ii} to be the probability that starting in state i, the process will reenter state i at some later time:

- If $P_{ii} = 1$, state i is called a *recurrent* or *absorbing* state.
- If $P_{ii} < 1$, state i is called a *transient* state.

Any state i is said to have period d if $P_{ij}^{n} = 0$ whenever n is not divisible by a number d such that d is the largest number with this property. Any state with a period 1 is called an *aperiodic* state. This means every state can recur at each time step.

An irreducible Markov chain consisting of aperiodic states is called an *irreducible* and *aperiodic Markov chain*.

1.13 The limiting probability

If a Markov chain is irreducible and aperiodic with transition probability P_{ij}^n, then the limit

$$q_j = \lim_{n \to \infty} P_{ij}^n \quad j \geq 0, \tag{1.75}$$

exists and is independent of the initial state (Ross 1989). q_j is an element of the *equilibrium* or *stationary steady-state* probability vector with the property that

$$q_j \geq 0, \quad \sum_j q_j = 1, \tag{1.76}$$

and it is the unique non-negative solution of the equation

$$q_j = \sum_i q_i P_{ij}, \quad j \geq 0. \tag{1.77}$$

We also note that

$$\sum_i q_j P_{ji} = q_j,$$

since $\mathbf{P}$ is a stochastic matrix. Thus we have

$$q_i P_{ij} = q_j P_{ji}. \tag{1.78}$$

Equation (1.75) states that after a large number of transitions, states will be distributed according to the equilibrium probability vector, which is independent of the initial state. The Markov chain must be irreducible if it is to have a unique *equilibrium distribution*; otherwise, the system may fall into a state from which it cannot enter some other states, and the system can no longer be independent of its initial configuration. All states of a system must also be aperiodic if there is to be a limiting equilibrium distribution; otherwise, the system will not exhibit a distribution of states that is independent of time.

Let us illustrate this concept with a simple example. Consider a process with only two states such that a single-step transition probability matrix is given by

$$\mathbf{P} = \begin{bmatrix} 0.6 & 0.4 \\ 0.3 & 0.7 \end{bmatrix}. \tag{1.79}$$

At step 2, we have

$$\mathbf{P}^{(2)} = \mathbf{PP} = \begin{bmatrix} 0.480 & 0.520 \\ 0.390 & 0.610 \end{bmatrix}. \tag{1.80}$$

Similarly, we have

$$\mathbf{P}^{(3)} = \begin{bmatrix} 0.433 & 0.566 \\ 0.425 & 0.574 \end{bmatrix},$$

$$\mathbf{P}^{(4)} = \begin{bmatrix} 0.4286 & 0.5713 \\ 0.4285 & 0.5714 \end{bmatrix},$$

$$\mathbf{P}^{(7)} = \begin{bmatrix} 0.4285 & 0.5714 \\ 0.4285 & 0.5714 \end{bmatrix},$$

$$\mathbf{P}^{(8)} = \begin{bmatrix} 0.4285 & 0.5714 \\ 0.4285 & 0.5714 \end{bmatrix}.$$

(1.81)

Note that after step 7, the transition probability matrix becomes nearly constant. The rows are nearly identical. This means that independent of whether the process is started with state 1 or state 2, the probability of occurrence of states 1 and 2 are, respectively, 0.4285 and 0.5714.

The SA methods described in Chapter 4 can be modeled using the concept of a Markov chain, and the proof of convergence of SA is based on the developments described in this chapter.

2

Direct, linear, and iterative-linear inverse methods

The goal of geophysics is to determine the properties of the earth's interior from the surface and/or boreholes using measurements of physical phenomena. In seismology, the data consist of seismograms (seismic wave amplitude as a function of time and distance) from earthquakes or man-made explosions. In electrical or magnetotelluric (MT) geophysics, the data are the measurements of apparent resistivity and phase at different spatial locations and periods. The problem, however, is common: *How do we determine subsurface structure and rock properties (elastic properties in the case of seismic data and electrical properties in the case of resistivity/ MT data) from these observations?* This constitutes an *inverse problem* – the problem of deriving properties of the earth or a *model* of the earth from the observed geophysical data that have been affected by the variation in properties of the earth material. First, to understand how the data are affected by the model, we must be able to calculate theoretical data for an assumed earth model. This constitutes the *forward problem*. This involves deriving a mathematical relationship between data and model. We think it safe to say that for many geophysical problems, the forward problems are fairly well understood in that the basic constituent equations have already been recognized. Most research has focused on finding solutions to these basic equations. In many cases, elegant analytic solutions can be obtained when some simplistic assumptions are made for the earth model. Unfortunately, the real earth is very complex, and it is the complexity or heterogeneity of the earth that is of interest to geophysicists. For complex earth models, analytic solutions are usually impossible to obtain, and a pure numerical or a combination of analytic and numerical techniques needs to be employed. Most numerical methods are based on finite difference or finite element formulations. With the advent of supercomputers and parallel computers, such methods are becoming increasingly popular. Descriptions of forward problems are beyond the scope of this book. We will only comment that for efficient use of inverse methods such as the ones that are described in this book, one must be well versed with the forward modeling algorithms. Knowledge of the forward modeling algorithms for the problems of interest is an essential element in formulating an efficient inversion algorithm based on global optimization techniques.

Geophysical inverse methods have been a subject of active research, and methods such as direct inversion and linear and iterative-linear methods have been studied in great detail. The global optimization methods forming the main subject of this book have only recently been applied to geophysical inverse problems. Although it may not be necessary to understand in detail the classical inverse methods, knowledge of the classical approaches will help the reader to better appreciate the effectiveness of these new methods. Also, it is necessary to recognize that many of the concepts developed in the more classical approaches may easily be extended to and even incorporated into the new techniques. Therefore, in this chapter we briefly describe these methods. Excellent texts on inverse theory as applied to geophysics are those by Menke (1984) and Tarantola (1987), and interested readers should refer to these texts for a detailed description of these methods. We summarize many of their results in this chapter so that notations are consistent with those in subsequent chapters. Also in our discussion we mostly draw examples from seismology/seismic exploration, although most techniques (except possibly the direct inversion methods) apply to other geophysical methods as well.

Geophysical inversion methods can be broadly classified into two categories: (1) direct inversion methods or operator-based inversion methods and (2) model-based inversion methods.

2.1 Direct inversion methods

Direct inversion methods are formulated based on the physics of the forward problem by recognizing or designing a mathematical operator that is applied to the observed data (sometimes recursively) to derive a model of the earth. Note that the model is derived from the data directly, and hence we prefer the name *direct inversion*. No assumptions are made on the values of the model parameters (such as compressional-wave velocity, density, etc.), but the variability of the model (whether it varies in one or two spatial dimensions, i.e., 1D or 2D) needs to be assumed. (Actually, no direct inversion method has been formulated for the general case of a heterogeneous anisotropic 3D medium.) One commonly used processing method in seismic exploration is called *migration*. Migration generates an image of the subsurface based on impedance contrasts directly from the data, but this operation cannot strictly be included in the category of direct inversion. This is so because migration requires that the subsurface velocity variation be known, and the algorithm uses this information to *undo* the propagation effects so as to place the reflected/diffracted energy at the spatial locations in the subsurface where they originated. It is generally assumed and has been demonstrated by many synthetic examples that in some situations one can obtain fairly good subsurface images even though the details of the velocity function may not be very well-known. However, as we will see in Chapter 7, the problem of migration velocity estimation can also be posed as an inverse problem.

The most well-known of all the direct inversion methods in seismology is called the *layer-stripping method* (Yagle and Levy 1983, 1985; Clarke 1984; Ziolkowski *et al.* 1989; Singh *et al.* 1989). The method is generally based on the reflectivity formulation (Fuchs and Müller 1971; Kennett 1983) of the forward problem in a 1D elastic (Clarke 1984) or acoustic (Ziolkowski *et al.* 1989) medium. In a medium in which the elastic properties vary only with depth, the elastic wave equation can be transformed into the frequency-wave number (ω, k) or frequency-ray parameter (ω, p) domain by the application of Fourier-Hankel transforms. This transformation reduces the wave equation into an ordinary differential equation in depth z that can be solved in either the (ω, k) or (ω, p) domain by propagator matrix methods. Forward modeling methods such as Thompson-Haskell or the reflectivity method (Fuchs and Müller 1971; Kennett 1983) are based on such principles. This results in an algorithm that computes the reflection response of a stack of layers in the (ω, p) domain by a recursive scheme over the number of layers. This, in essence, gives the plane-wave response $R(\omega, p)$ of the medium at each frequency. The responses of these plane waves are summed properly, and an inverse Fourier transform is applied to obtain synthetic seismograms in the offset-time (x, t) domain for a plane-layered earth. The same principle can be applied in the reverse order to determine rock properties from measured seismic traces. The first step of such an inversion scheme would be to transform the data from (x, t) to the intercept time-ray parameter ($\tau - p$) domain, i.e., to derive the plane-wave response from the recorded data by a procedure called a *plane-wave decomposition* (e.g., Brysk and McCowan 1986). Then successively remove the effect of each layer by downward continuation (layer stripping) while deriving the properties of each layer.

We illustrate the layer-stripping method for a horizontally layered acoustic medium following Ziolkowski *et al.* (1989). Assume that

ρ_i = density of the *i*th layer;
α_i = compressional-wave velocity of the *i*th layer;
h_i = thickness of the *i*th layer, $i = 0, 1, 2, ..., N$;

where layers 0 and $N + 1$ represent the two half spaces (Figure 2.1). For a plane wave incident at an angle θ_i, with horizontal slowness $p = \sin \theta_i/\alpha_i$ and vertical

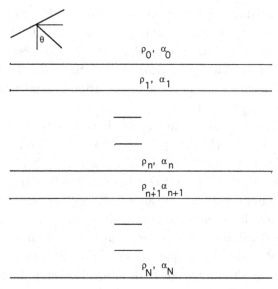

Figure 2.1 An acoustic layered model used in the formulation of direct inversion by the layer-stripping method.

slowness $q_i = \cos \theta_i / \alpha_i$, the acoustic plane-wave reflection coefficient for the ith layer is given by

$$\Gamma_i = \frac{(q_i / \rho_i) - (q_{i+1} / \rho_{i+1})}{(q_i / \rho_i) + (q_{i+1} / \rho_{i+1})}, \tag{2.1}$$

and the vertical delay time is given by

$$\tau_i = 2 q_i h_i, \quad i = 1, 2, \dots, N. \tag{2.2}$$

The global reflection response can be computed by recursively computing the phase delay (vertical component of travel time) and amplitude (reflection coefficients) through the sequence of layers. This is obtained by the following recursion formula:

$$R_i(\omega, p) = \frac{\Gamma_i(\omega, p) + R_{i+1}(\omega, p) \exp(i\omega\tau_{i+1})}{1 + \Gamma_i(\omega, p) R_{i+1}(\omega, p) \exp(i\omega\tau_{i+1})}. \tag{2.3}$$

Note that Γ_i can, in general, be a function of frequency and slowness, e.g., when the velocities become frequency dependent. Similar expressions for an elastic medium require a matrix formulation (e.g., Kennett 1983). Equation (2.3) is a mathematical description that can be used to describe deconvolved field data after they have been properly plane-wave decomposed. For example, the plane-wave response of a layered medium can be computed by calculating the response of each layer in sequence from bottom to top (layers $i = N - 1, N - 2, ..., 1$) such that $R_1(\omega, p)$ is the response of the medium. To invert plane-wave-decomposed data for earth model parameters, we can use Eq. (2.3) in reverse order, i.e., by taking out the effect of each layer from the data in sequence from the top to the bottom. This requires rewriting Eq. (2.3) in the following form:

$$R_{i+1}(\omega, p) \exp(i\omega\tau_{i+1}) = \frac{\Gamma_i - R_i(\omega, p)}{\Gamma_i R_i(\omega, p) - 1}$$
$$= \Gamma_i + X_i(\omega, p),$$

(2.4)

where

$$x_i(\omega, p) = \frac{(1 - \Gamma_i^2) R_i(\omega, p)}{1 - \Gamma_i R_i(\omega, p)}.$$

(2.5)

Note that the plane-wave-decomposed data are represented by $R_1(\omega, p)$. Equation (2.4) can now be used recursively to strip off layers to obtain $R_2(\omega, p)$, $R_3(\omega, p)$, ..., $R_N(\omega, p)$. From these equations and using some prior information, the basic acoustic parameters of the layers (acoustic impedance, compressional-wave velocity, and density) can be computed. An example of inversion using this method is shown in Figure 2.2 (taken from Ziolkowski *et al.* 1989). Note that several authors used the layer-stripping method in seismic inversion. We chose to describe the algorithm based on the work of Ziolkowski *et al.* (1989) primarily because of the simplicity with which these authors described the technique.

The theory of layer stripping is based on the assumption that the data are broadband and noise-free. However, seismic data are recorded only in a finite bandwidth of frequencies; i.e., some of the low-frequency information and high-frequency information are missing from the data. Therefore, for stable estimates of model parameters, this information will have to be supplied independently. Also, any noise present in the data will cause errors in the estimation of reflection coefficients. These errors grow rapidly and accumulate with depth (because of recursion), and the algorithm becomes unstable. Although some attempts have been made (e.g., Ziolkowski *et al.* 1989; Singh *et al.* 1989) to circumvent these problems, they are far from being satisfactory in practical situations where the data are band limited and contaminated with noise.

A layer-stripping type of approach can be formulated for any problem for which a recursive type forward modeling formulation can be written down. An example of a layer-stripping method applied to MT data can be found in Levy *et al.* (1988).

One other method that can be included in the category of direct inversion is based on an inverse scattering formulation or Born theory popularly known as *Born inversion* (e.g., Clayton and Stolt 1981; Weglein and Gray 1983; Stolt and Weglein 1985). In this approach, the earth is assumed to vary in 2D or 3D, relaxing

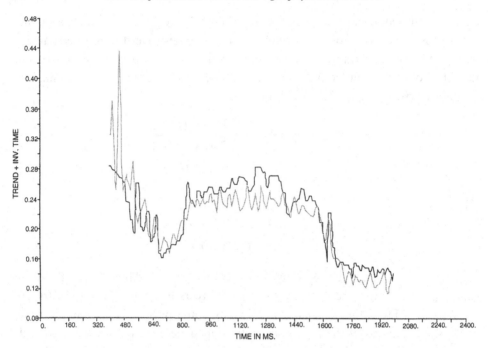

Figure 2.2 An example of direct inversion of real seismic data by layer strip-ping. Admittance profiles are shown. Solid line corresponds to the profile derived from well logs. The dashed line corresponds to the admittance profile obtained by layer-stripping inversion. (From Ziolkowski *et al.* 1989.)

the most fundamental restriction of the layer-stripping approach. The popularity of the method results from the fact that Born inversion implicitly includes migra-tion (imaging) with the proper recovery of band-limited material property (or their contrast) estimates. The slowly varying or *background* velocity must be provided to control the imaging process, although Born inversion to recover the background velocity could be used as part of a more general model-based inversion approach. However, to remove the effect of multiples, the process must be applied iteratively. This is not routinely done, and the method suffers from the same limitations as the layer-stripping method because the data are band limited and contain noise.

Any method of inversion will suffer when incomplete data contaminated with noise are used to derive material property estimates. Direct or operator-based inver-sion methods are particularly sensitive to these problems because it is often not obvious how to condition the operator to minimize the deleterious effects caused by the application of these operators to real data. These problems are somewhat easier to overcome in model-based inversion where uncertainties in the data can be incorporated into the inversion scheme.

2.2 Model-based inversion methods

Unlike direct inversion methods, we make no attempt to reverse the forward process in a model-based inversion approach. In this approach, synthetic data are
generated for an assumed model and compared with observed data. If the match
between observed and synthetic data is acceptable, the model is accepted as the
solution. Otherwise, the model is changed; the synthetics are recomputed and compared again with the observations. This iterative forward modeling procedure is
repeated until an acceptable match is obtained between data and synthetics. Thus,
in this approach, inversion is viewed as an *optimization* process in which a model
is sought that best explains the observations. The fitness (measure of agreement)
or misfit (measure of disagreement) between observed and synthetic data is used
as a measure of acceptability of an earth model. Optimization methods vary based
on their method of search for the optimal model. The simplest of the optimization methods is one that assumes a linear relationship between data and model, in
which case an answer can be obtained in one step.

In geophysics, we always deal with discrete data, and thus it is convenient to
represent the data as a column vector of the type

$$\mathbf{d} = \left[d_1, d_2, d_3, \ldots, d_{ND} \right]^T,\tag{2.6}$$

where *ND* is the number of data points and *T* is a matrix transpose. For example, in
seismology each data value is a time sample in a seismogram, and $ND = NT \bullet NTR$,
where *NT* is the number of samples in a seismogram and *NTR* is the number of
seismograms. In direct-current resistivity sounding problems, each d_i corresponds
to an apparent resistivity value measured at a given electrode separation.

Similarly, an earth model can also be represented by a column vector

$$\mathbf{m} = \left[m_1, m_2, m_3, \ldots, m_{NM} \right]^T,\tag{2.7}$$

where *NM* is the total number of model parameters. For example, for a 1D acoustic
problem in seismology, as described in the preceding section, the model parameters can be used to represent compressional-wave velocity (α_i), density (ρ_i), and
thickness (h_i) of all the layers.

Synthetic data are generated by forward calculation using a model vector **m**.
Thus the synthetic data vector $\mathbf{d}_{\text{syn}}$ can be computed by application of a *forward
modeling operator* g to the model vector **m**; i.e.,

$$\mathbf{d}_{\text{syn}} = g(\mathbf{m}).\tag{2.8}$$

In most geophysical problems, the forward modeling operator g is a non-linear operator. One example of this is given in Section 2.1 (Eq. 2.3) for a 1D seismic (acoustic) case in which the reflection response (data) is given by a non-linear function of model parameters – compressional-wave velocity, density, and thickness.

The inverse problem now reduces to determining model(s) that minimize(s) the misfit between observed and synthetic data. The misfit function is also called the *objective function, cost function, error function, energy function*, etc. and is usually given by a suitably defined *norm*. If the error vector **e** is given by

$$\mathbf{e} = \mathbf{e}_{obs} - \mathbf{d}_{syn} = \mathbf{d}_{obs} - g(\mathbf{m}), \tag{2.9}$$

a general norm L_p (e.g., Menke 1984) is defined as

$$L_p \text{ norm:} \qquad \| \mathbf{e} \|_p = \left[\sum_{i-1}^{ND} |e_i|^p \right]^{1/p}, \tag{2.10}$$

where *ND* is the number of data points. Thus the commonly used L_2 norm is given by

$$L_2 \text{ norm:} \qquad \| \mathbf{e} \|_2 = \left[\sum_{i-1}^{ND} |e_i|^2 \right]^{1/2}, \tag{2.11}$$

or in vector notation we have

$$\| \mathbf{e} \|_2 = \left[\left(\mathbf{d}_{obs} - g(\mathbf{m}) \right)^T \left(\mathbf{d}_{obs} - g(\mathbf{m}) \right) \right]. \tag{2.12}$$

Usually this equation is divided by the number of observation points *ND*, in which case it reduces to the well-known root mean square (RMS) error. Higher norms give larger weights to large elements of **e**. In many geophysical applications, the L_2 norm has been used; use of other types of norms, e.g., L_1 (Claerbout and Muir 1973) can also be found in the geophysical literature.

At this point we note the following features of model-based geophysical inverse problems:

- Since g in Eq. (2.8) is in general a non-linear operator, the error function (as a function of the model) is complicated in shape. Thus it can be expected to have multiple minima of varying heights. Only if g actually is or can be approximated by a linear operator will the error function become quadratic with respect to perturbations in the model parameters (this will be elaborated later in this chapter) and will have only one well-defined minimum.
- For many geophysical applications, generation of synthetic data is a highly time-consuming task even on very fast computers.

- If each model parameter can take M possible discrete values and there are N model parameters, then there are M^N possible models to be tested. Typically, model spaces of the order of 50^{50} or higher are common.

Depending on the search method applied to find optimal solutions, model-based inversion methods can be classified into the following categories.

2.2.1 Linear/linearized methods

These methods assume that the data are a linear function of the model parameters. Then it is possible to replace the forward modeling operator $\mathcal{g}$ in Eq. (2.8) with a linear operator, a matrix $\mathbf{G}$, to obtain

$$\mathbf{d} = \mathbf{Gm}. \tag{2.13}$$

This inverse problem can now be solved with the well-known methods of linear algebra.

2.2.2 Iterative-linear or gradient-based methods

These methods can also be called *calculus-based methods* because they use gradient (derivative) information to compute an update to the current model. The error between the data and synthetics for the current model is mapped into model updates, and the procedure is repeated until the updates become very small.

2.2.3 Enumerative or grid-search method

This method involves searching through each point in model space. Computation of synthetic data for a large model space for many geophysical applications is a formidable task and not usually a practical solution.

2.2.4 Monte Carlo method

This method involves the random sampling of model space. It is hoped that in some small and finite number of trials, we may be able to obtain good answers. This is a completely blind search and therefore may be very computationally expensive.

2.2.4.1 Directed Monte Carlo methods

The global optimization methods – simulated annealing (SA) and genetic algorithms (GAs) belong to this category. These methods use random sampling with some directivity to guide their search.

2.3 Linear/linearized inverse methods

Linear inverse methods are the best known of all geophysical methods and have been in use for decades. The basic assumption for this formulation is that the data are a linear function of the model parameters, and naturally, this approach will work best where such assumptions are valid. In many cases we may be able to linearize the relationship between data and model when certain conditions are applied. For example, the reflection coefficients as given by the Zoeppritz equations (Aki and Richards 1980) are a non-linear function of the compressional-wave velocity, shear-wave velocity, and density. However, a linearized form of these equations can be derived that is valid for small angles of reflection. Another approach to linearization is to obtain a linear relationship between perturbations in the data and perturbations in the model. This is achieved as follows: Assume that the observed data $(\mathbf{d}_{obs})$ may be obtained by the action of the forward modeling operator g on a model obtained by applying a small perturbation $\Delta\mathbf{m}$ to a reference model $\mathbf{m}_0$. The data obtained from the reference model $\mathbf{m}_0$ will be called *synthetic data* $\mathbf{d}_{syn.}$ That is,

$$\mathbf{d}_{obs} = g\left(\mathbf{m}_0 + \Delta\mathbf{m}\right), \tag{2.14}$$

and

$$\mathbf{d}_{syn} = g\left(\mathbf{m}_0\right). \tag{2.15}$$

We now expand Eq. (2.14) in a Taylor's series around the reference model $\mathbf{m}_0$ to obtain

$$\mathbf{g}\left(\mathbf{m}_0 + \Delta\mathbf{m}\right) = \mathbf{g}\left(\mathbf{m}_0\right) + \left.\frac{\partial\mathbf{g}\left(\mathbf{m}_0\right)}{\partial\mathbf{m}}\right|_{m=m_0} \Delta\mathbf{m} \tag{2.16}$$

+ second and higher-order terms.

Ignoring the higher-order terms and using Eqs. (2.14) and (2.15), we obtain

$$\mathbf{d}_{obs} = \mathbf{d}_{syn} + \left.\frac{\partial\mathbf{g}\left(\mathbf{m}_0\right)}{\partial\mathbf{m}}\right|_{m=m_0} \Delta\mathbf{m} \tag{2.17}$$

or

$$\Delta\mathbf{d} = \mathbf{G}_0\Delta\mathbf{m}, \tag{2.18}$$

where $\Delta\mathbf{d} = \mathbf{d}_{obs} - \mathbf{d}_{syn}$, and $\mathbf{G}_0$ is the sensitivity matrix or the matrix of partial derivatives of synthetic data with respect to model parameters. We now have a linear relationship between perturbations in data and perturbations in the model parameters.

As an example, consider the problem of seismic tomography in which we attempt to model travel-time observations by computing the travel time for ray paths from the source to the receivers. If $\alpha(x_i, y_i, z_i)$ is the compressional-wave velocity at a spatial location (x_i, y_i, z_i), the travel time for a ray path is given by the following sum:

$$T = \sum_{i \in S} \frac{ds_i}{\alpha(x_i, y_i, z_i)} = \sum_{i \in S} U(x_i, y_i, z_i) ds_i, \qquad (2.19)$$

where S denotes a ray path between source and receiver and U is the reciprocal of the compressional-wave velocity α, called *slowness*. This equation is non-linear because the ray path S is itself a function of the unknowns $U(x, y, z)$ – the model parameters for this problem. When the inverse problem is represented by Eq. (2.18), the data residual vector now represents the travel-time residuals (difference between the measured travel times and the travel time computed for a reference model). The model perturbation vector gives the deviation from the reference model, and the matrix of derivatives contains the partial derivatives of travel time with respect to slowness values computed for the reference model.

In the tomography problem, we can exploit *Fermat's principle*, which states that the travel time is stationary with respect to first-order changes in the ray path. Therefore, for small perturbations in slowness values, no ray-path calculation is necessary, and the perturbations in travel time can be calculated using the old ray paths but with the new perturbed slowness values. This makes calculation of the matrix of derivatives very rapid. Note, however, that this assumption is valid only for small perturbations in the model.

Once we recognize that a problem is linear or linearizable with respect to a reference model, equations of the form (2.13) and (2.18) can be written down. In Eq. (2.8), the matrix $\mathbf{G}$ represents a forward modeling operator and is known from the forward relation. In Eq. (2.18), the elements of the $\mathbf{G}_0$ matrix are the partial derivatives of data with respect to model parameters (computed for the reference model), which need to be calculated. Once $\mathbf{G}$ is known, we can solve for the model vector by applying the inverse operator for $\mathbf{G}$, $\mathbf{G}^{-1}$ (if it exists), to the data vector. Similarly applying the inverse of $\mathbf{G}_0$ to the data residual vector, we can obtain an update to the model perturbation vector. This, however, may not be a trivial task, and much of the subject of linear inverse theory deals with issues related to obtaining estimates of the inverse operators for the solution of linear inverse problems. The solution methods, the nature of the solution obtained, and the conditions under which solutions may be obtained are but a few of the topics where geophysical inverse theorists were able to introduce many new ideas and concepts (e.g., Backus 1970; Backus and Gilbert 1967, 1968, 1970; Jackson 1972, 1979; Parker 1972, 1977; etc.).

Formally, the inverse solution of Eq. (2.13) can be written as

$$\mathbf{m} = \mathbf{G}^{-1}\mathbf{d}, \tag{2.20}$$

where $\mathbf{G}^{-1}$ is the inverse operator of $\mathbf{G}$. This leads us to the following issues related to the problem of finding solutions (e.g., Koch 1990; Tarantola 1987).

2.3.1 Existence

The existence implies that of a solution $\mathbf{m}$ or of the inverse operator $\mathbf{G}^{-1}$.

2.3.2 Uniqueness

A solution is said to be *unique* if in changing a model from $\mathbf{m}_1$ to $\mathbf{m}_2$, the data also change from $\mathbf{d}_1$ to $\mathbf{d}_2$ such that $\mathbf{d}_1 \neq \mathbf{d}_2$; i.e., the operator $\mathbf{G}$ is *injective*. Otherwise, several models will explain the data equally well, and we obtain non-unique solutions. Some of the reasons for the non-uniqueness of the results for geophysical inverse problems are

- For problems dealing with earth parameters, the infinite dimensionality of the model space is usually formulated into a discrete finite problem resulting in inherent non-uniqueness (e.g., Backus and Gilbert 1967). Material properties in the real earth are continuous functions of the spatial coordinates. In inversion, attempts are made to derive these functions from measurements of a finite number of data points, causing non-uniqueness in results.
- Uniqueness is also related to the problem of *identifiability* of the model by the data. For example, in seismic problems, the information on seismic wave velocity is contained in the move-out (travel time as a function of offset) in the data. Therefore, the use of only near offset traces does not reveal much information on velocity. This means that a suite of velocities will explain the data equally well, resulting in non-unique estimates of velocities. In seismic tomography problems, the regions of the earth with no ray coverage cannot be resolved from the data, and slowness estimates for these regions will result in non-unique values. Thus resolution and uniqueness are closely related. Also, in many situations we may have a clear tradeoff between model parameters.

2.3.3 Stability

Stability indicates how small errors in the data propagate into the model. A stable solution is insensitive to small errors in the data values. Instability may induce non-uniqueness, which enhances the inherent non-uniqueness in geophysical applications.

2.3.4 Robustness

Robustness indicates the level of insensitivity with respect to a small number of large errors (outliers) in the data. Inverse problems that do not possess uniqueness and stability are called *ill-posed* inverse problems. Otherwise, the inverse problem is called *well-posed*. Techniques known as *regularization* can be applied to ill-posed problems to restore well-posedness (Koch 1990).

Since many geophysical inversions result in non-unique solutions, the objective of inversion is to first find a solution (or solutions) and to represent the degree of non-uniqueness of the solution in a quantitative manner. During the process, attempts should be made to reduce non-uniqueness, if possible, and/or to explain it in terms of data errors and the physics of the forward problem.

Now we discuss methods of solution of inverse problems starting with the linear inverse problem. In all our discussions that follow in this chapter, we assume that ND = the number of data points and NM = the number of model parameters unless otherwise noted.

2.4 Solution of linear inverse problems

2.4.1 Method of least squares

Recall that for a linear inverse problem (Eq. 2.13), the L_2-norm error function is given by

$$E(\mathbf{m}) = \mathbf{e}^T \mathbf{e} = (\mathbf{d} - \mathbf{Gm})^T (\mathbf{d} - \mathbf{Gm}). \tag{2.21}$$

This can be solved by locating the minimum of E where the derivative of E (with respect to $\mathbf{m}$) vanishes. That is,

$$\frac{\partial E(\mathbf{m})}{\partial \mathbf{m}} = 0, \tag{2.22}$$

which gives

$$\mathbf{G}^T \mathbf{Gm} - \mathbf{G}^T \mathbf{d} = 0, \tag{2.23}$$

where $\mathbf{0}$ is a null vector. This yields

$$\mathbf{m}_{\text{est}} = \left[\mathbf{G}^T \mathbf{G}\right]^{-1} \mathbf{G}^T \mathbf{d}, \tag{2.24}$$

assuming that $[\mathbf{G}^T\mathbf{G}]^{-1}$ exists, where $\mathbf{m}_{est}$ is the least squares estimate of the model. Thus, from knowledge of the $\mathbf{G}$ matrix, one can derive estimates of the model parameters in one step. $\mathbf{G}$ is an $(ND \times NM)$ matrix, and $\mathbf{G}^T\mathbf{G}$ is a symmetric square matrix with $(NM \times NM)$ elements. Whether the least squares solution exists or not depends on $[\mathbf{G}^T\mathbf{G}]^{-1}$, which, in turn, depends on how much information the data vector $\mathbf{d}$ possesses on the model parameters. The matrix $[\mathbf{G}^T\mathbf{G}]^{-1}\mathbf{G}^T$ operates on the data to derive model parameters; i.e., it inverts the system of linear equations (Eq. 2.13). We will represent this matrix with the symbol $\mathbf{G}^g$ such that

$$\mathbf{G}^g = \left[\mathbf{G}^T\mathbf{G}\right]^{-1}\mathbf{G}^T. \tag{2.25}$$

One of the necessary conditions for the system of linear equations $\mathbf{Gm} = \mathbf{d}$ to have one unique solution is that there are as many equations as the number of unknown model parameters. This is the situation when we have as many data as the number of model parameters ($ND = NM$) such that these data contain information on all the model parameters. Such a problem is called an *even-determined problem*. In an *underdetermined problem*, $NM > ND$; i.e., there are fewer data than the number of model parameters. This would usually be the case with geophysical inversion if we attempt to estimate earth model parameters that are continuous (or infinite) from a finite set of measurements. The problem can be reduced to an *even-determined* or even an *overdetermined* case by discretization that reduces the number of model parameters. For example, instead of solving for the earth parameters as a continuous function of spatial coordinates, we may divide the earth model into a set of discrete layers (based on independent information). A problem may be *mixed-determined* as well, in which data may contain complete information on some model parameters and none on the others. Such a situation may arise in seismic tomography problems when there is complete ray coverage in some of the blocks, whereas other blocks may be completely devoid of any ray coverage. Thus a problem can be underdetermined even in cases with $ND > NM$. In the *underdetermined* case, several solutions exist, and the model parameter estimates will be non-unique. In situations where we have more data than the number of model parameters and the data contain information on all model parameters, the problem is said to be *overdetermined*. Clearly, in such a situation the linear system of equations cannot find an answer that fits all data points unless they all lie on a straight line. In this case the best estimate can be obtained in the least squares sense, meaning that the error is a non-zero smallest value.

2.4.1.1 *Maximum-likelihood methods*

Geophysical data are often contaminated with noise, and therefore, every data point may be uncertain, the degree of uncertainty being different for different data points. It is also possible that the data points may influence each other; i.e., they may be correlated. Thus, based on the theory outlined in Chapter 1, each data point can be considered a random variable such that the vector **d** now represents a vector of random variables. If each of the data variables is assumed to be Gaussian distributed, their joint distribution is given by

$$p(\mathbf{d}) \propto \exp\left[-\frac{1}{2}(\mathbf{d}-\langle\mathbf{d}\rangle)^T C_d^{-1}(\mathbf{d}-\langle\mathbf{d}\rangle)\right], \tag{2.26}$$

where <d> and C_d are the mean data and data covariance matrix, respectively. $P(\mathbf{d})$ gives the probability of observation of the data values. For a linear inverse problem, we can write the preceding probability as

$$p(\mathbf{d}) \propto \exp\left[-\frac{1}{2}(\mathbf{d}-G\mathbf{m})^T C_d^{-1}(\mathbf{d}-G\mathbf{m})\right], \tag{2.27}$$

This means that the application of **G** on the estimated model results in the mean values of data and that **d** is a single realization of the random data vector. Thus the optimal values of the model parameters are the ones that maximize the probability of measured data, given the uncertainties in the data. This procedure is an example of the *maximum-likelihood method* (MLM). The use of the term MLM will be apparent in the context of Bayesian formulation (discussed later in Section 2.6), in which case Eq. (2.27) represents the likelihood function. Note that the MLM does not necessarily imply a Gaussian likelihood function as given by Eq. (2.27).

For numerical implementation, this method is not very different from the least squares method (for the case of Gaussian-distributed data) described in the preceding section. The maximum probability occurs when the argument of the exponential is minimum. Thus, instead of minimizing the error function

$$E(\mathbf{m}) = (\mathbf{d}-G\mathbf{m})^T(\mathbf{d}-G\mathbf{m}), \tag{2.28}$$

we minimize the following error function:

$$E_1(\mathbf{m}) = (\mathbf{d}-G\mathbf{m})^T C_d^{-1}(\mathbf{d}-G\mathbf{m}), \tag{2.29}$$

which can again be solved by the least squares method. By finding the minimum where the derivative of the error with respect to the model parameters is zero, we can derive an equation for $\mathbf{m}_{est}$ analogous to Eq. (2.24). This form reduces to the case of simple least squares when C_d is an identity matrix. In the general case, however, each data residual is weighted inversely with the corresponding element of the data covariance matrix. This means that for a diagonal C_d, the data with large uncertainty will have a relatively smaller contribution to the solution,

whereas those with small error will have a relatively larger contribution. The process attempts to fit the reliable data values better than the unreliable ones.

Even without knowledge of C_d as required in the MLM, it may be desirable to
use a weighting matrix W_d as in the following definition of the error function:

$$E_2(\mathbf{m}) = (\mathbf{d} - G\mathbf{m})^T \mathbf{W}_d (\mathbf{d} - G\mathbf{m}), \tag{2.30}$$

The need for a weighting matrix can occur due to assumed data uncertainties or
because the elements of the data vector may have different dimensions or ranges of
values. In the conventional definition of the error function (Eq. 2.28), all the data
values have equal weight, and therefore, the data with large numerical values will
dominate. One way of weighing is to normalize the data residuals by the observed
data. Many different weighing schemes can and have been developed for different
applications.

2.4.2 Stability and uniqueness – singular-value-decomposition (SVD) analysis

For the system of linear equations of the type $\mathbf{d} = \mathbf{Gm}$, the solution can be obtained
by the method of least squares (Eq. 2.24). This involves forming a square matrix
$\mathbf{G}^T\mathbf{G}$ that is a non-negative definite square symmetric matrix that can be solved by
Cholesky decomposition so that no explicit matrix inversions need to be computed
(Lines and Treitel 1984). There may, however, be some numerical inaccuracies in
generating $\mathbf{G}^T\mathbf{G}$, in which case it may be desirable to deal directly with the system
$\mathbf{d} = \mathbf{Gm}$ with solution $\mathbf{m}_{\text{est}} = \mathbf{G}^{-1}\mathbf{d}$. In general, $\mathbf{G}$ is not a square matrix; therefore,
it is not possible to determine its inverse by conventional approaches. The SVD
of the matrix $\mathbf{G}$ is a powerful technique for computing the inverse of $\mathbf{G}$ (Lanczos
1961). The SVD of a matrix $\mathbf{G}(ND \infty NM)$ with rank[1] $r(G) = p$, $(\max(p) = \min(ND,$
$NM))$ can be written as

$$\mathbf{G} = \mathbf{U} \Lambda \mathbf{V}^T, \tag{2.31}$$

where $\mathbf{U}$ is an $ND \times ND$ square orthogonal matrix ($\mathbf{U}^T = \mathbf{U}^{-1}$) and $\mathbf{V}$ is an $NM \times NM$
orthogonal matrix ($\mathbf{V}^T = \mathbf{V}^{-1}$). Note, however, that the inverses of $\mathbf{U}$ and $\mathbf{V}$ are often
not defined. It only holds that $\mathbf{U}^T\mathbf{U} = \mathbf{I}$ and $\mathbf{V}^T\mathbf{V} = \mathbf{I}$. Λ denotes a matrix whose
diagonal elements are called *singular values*. As an example for $ND = 3$, we write
the SVD of $\mathbf{G}$ (assuming that $ND = 3$ and $NM = 4$, i.e., the underdetermined case)
as follows:

---

[1] The minor of a matrix $\mathbf{A}$ is defined to be the determinant of a square submatrix $\mathbf{A}$. The rank of a matrix $\mathbf{A}$, $r(\mathbf{A})$,
is the order of the largest non-zero minor of $\mathbf{A}$.

$$\begin{bmatrix} G_{11} & G_{12} & G_{13} & G_{14} \\ G_{21} & G_{22} & G_{23} & G_{24} \\ G_{31} & G_{32} & G_{33} & G_{34} \end{bmatrix} = \begin{bmatrix} U_{11} & U_{12} & U_{13} \\ U_{21} & U_{22} & U_{23} \\ U_{31} & U_{32} & U_{33} \end{bmatrix} \begin{bmatrix} \Lambda_{11} & 0 & 0 & 0 \\ 0 & \Lambda_{22} & 0 & 0 \\ 0 & 0 & \Lambda_{33} & 0 \end{bmatrix}$$

$$\begin{bmatrix} V_{11} & V_{12} & V_{13} & V_{14} \\ V_{21} & V_{22} & V_{23} & V_{24} \\ V_{31} & V_{32} & V_{33} & V_{34} \\ V_{41} & V_{42} & V_{43} & V_{44} \end{bmatrix}^{T} . \tag{2.32}$$

We also assume that the Λ_{ii} are arranged in decreasing order. It turns out that the eigenvector matrices $\mathbf{V}$ and $\mathbf{U}$ can each be separated into two submatrices

$$\mathbf{V} = \begin{bmatrix} \mathbf{V}_p | \mathbf{V}_0 \end{bmatrix}, \tag{2.33}$$

and

$$\mathbf{U} = \begin{bmatrix} \mathbf{U}_p | \mathbf{U}_0 \end{bmatrix}, \tag{2.34}$$

where the matrices $\mathbf{V}_p$ and $\mathbf{U}_p$ are associated with singular values $\Lambda_i \neq 0$, and the matrices $\mathbf{V}_0$ and $\mathbf{U}_0$ are associated with the singular values $\Lambda_i = 0$. The $\mathbf{V}_0$ and $\mathbf{U}_0$ spaces are blind spots not illuminated by the operator $\mathbf{G}$ and are also called *null spaces*. For example, in seismic deconvolution, the null space corresponds to the frequencies that fall outside the estimated wavelet passband (Deeming 1987). The null space of the residual statics correction problem corresponds to the wavelengths greater than the recording cable length (Wiggins *et al.* 1976). The null space in cross-well seismic tomography corresponds primarily to rapid horizontal velocity variations. The null space can be caused by the problem being underdetermined. It can also be due to the identifiability of part of the model by the data. Clearly, in our earlier example, $ND = 3$ and $NM = 4$, the problem is underdetermined, and $\mathbf{V}$ has null space. We can rewrite the $\mathbf{G}$ matrix as follows:

$$\mathbf{G} = \begin{bmatrix} \mathbf{U}_p | \mathbf{U}_0 \end{bmatrix} \begin{bmatrix} \Lambda_p & 0 \\ 0 & 0 \end{bmatrix} \begin{bmatrix} \mathbf{V}_p | \mathbf{V}_0 \end{bmatrix}^{T} . \tag{2.35}$$

We find that $\mathbf{U}_p \approx \mathbf{U}_0$ spans up the data space, whereas $\mathbf{V}_p \approx \mathbf{V}_0$ spans up the model space. Thus the eigenvector matrices $\mathbf{V}_p$ and $\mathbf{V}_0$ determine the uniqueness of the solution, whereas $\mathbf{U}_p$ and $\mathbf{U}_0$ determine the existence of the solution. We can rewrite the preceding equation as

$$\mathbf{G} = \mathbf{U}_p \Lambda_p \mathbf{V}_p^{T},$$
$$\mathbf{G}^{-1} = \mathbf{V}_p \Lambda_p^{-1} \mathbf{U}_p^{T}. \tag{2.36}$$

Now we have

$$\mathbf{m}_{est} = \mathbf{G}^{-1}\mathbf{d} = \mathbf{G}^{-1}\left[\mathbf{Gm}\right] = \mathbf{V}_p\Lambda_p^{-1}\mathbf{U}_p^T\mathbf{U}_p\Lambda_p\mathbf{V}_p^T\mathbf{m} = \left[\mathbf{V}_p\mathbf{V}_p^T\right]\mathbf{m} = \mathbf{Rm},\qquad(2.37)$$

where $\mathbf{R} = \mathbf{V}_p\,\mathbf{R} = \mathbf{V}_p\,\mathbf{V}_p^T$ Note that if the null space for $\mathbf{V}$, $\mathbf{V}_0 = 0$, then $\mathbf{R} = \mathbf{VV}^T = \mathbf{I}$, an identity matrix, and the estimated model will be the same as the true model. Otherwise, the estimated model parameters will be weighted averages of the true model parameters, with the weights being given by the elements of the matrix $\mathbf{R}$, also called the *resolution matrix*.

Similarly, we also have

$$\mathbf{d}_{est} = \mathbf{Gm}_{est} = \left[\mathbf{U}_p\mathbf{U}_p^T\right]\mathbf{d} = \mathbf{Nd}.\qquad(2.38)$$

where $\mathbf{N} = \mathbf{U}_p\,\mathbf{U}_p^T$. Also, if the null space for $\mathbf{U}$, $\mathbf{U}_0 = 0$, then $\mathbf{N} = \mathbf{I}$ and the estimated data will be exactly the same as the observed data. The matrix $\mathbf{N}$ is called the *data-importance matrix*.

2.4.3 Methods of constraining the solution

As discussed earlier, the geophysical inverse problem is inherently ill-posed, resulting in non-unique estimates of earth model parameters. The method known as the *Backus–Gilbert approach* seeks average models to represent the solution of this problem. See Aki and Richards (1980), Menke (1984), and Tarantola (1987) for a discussion of this approach.

Often the large number of unknowns can be approximated by a small number of coarse grid points (e.g., the earth is assumed to be made up of a small number of discrete layers), which imposes well-posedness of the inverse problem. This constraint may be done independent of the data $\mathbf{d}$ and is called *prior* (or *a priori*) information or assumptions (Jackson 1979). Prior information can take many different forms (e.g., Jackson 1979). We describe some of them below.

2.4.3.1 Positivity constraint

Based on the physics of the problem, we may constrain the model parameters not to have any negative values. For example, seismic wave velocities and densities are always positive numbers, and the inversion results may be constrained to have only positive values for these parameters in seismic waveform inversion.

2.4.3.2 Prior model

In many instances we may have a very good reference (or prior) model. Thus in the inversion, in addition to minimizing the misfit between observed and synthetic data, we impose the restriction that the model does not deviate much from the prior model $\mathbf{m}_p$. This can be done by modifying the error function $E(\mathbf{m})$ to the following form:

$$E_3(\mathbf{m}) = (\mathbf{d} - \mathbf{Gm})^T (\mathbf{d} - \mathbf{Gm}) + \varepsilon (\mathbf{m} - \mathbf{m}_p)^T (\mathbf{m} - \mathbf{m}_p), \qquad (2.39)$$

where ε is a weight that decides the relative importance between model and data error.

In general, this weight can be different for different model parameters. This is so because, in some situations, we may have very good information on some model parameters and we do not want them to vary significantly, whereas other model parameters may not be so well-known in advance and must be allowed to deviate from $\mathbf{m}_p$. A convenient way of accomplishing this weighting is by employing a model covariance matrix $\mathbf{C}_m$ (assuming that the prior pdf of the model is Gaussian), which gives a quantitative estimate of the certainty we have on the prior model. Thus the error function $E_1(\mathbf{m})$ (Eq. 2.29) used in the MLM can be modified to the following:

$$E_4(\mathbf{m}) = (\mathbf{d} - \mathbf{Gm})^T \mathbf{C}_d^{-1} (\mathbf{d} - \mathbf{Gm}) + (\mathbf{m} - \mathbf{m}_p)^T \mathbf{C}_m^{-1} (\mathbf{m} - \mathbf{m}_p). \qquad (2.40)$$

2.4.3.3 Model smoothness

In many situations we may require that the model be smooth, i.e., slowly varying in some direction. The smoothness of the model can be imposed by means of a model weighting matrix $\mathbf{W}_m$ (which may have different forms based on the degree of smoothness desired), and the error function $E_2(\mathbf{m})$ can be modified to the following form:

$$E_5(\mathbf{m}) = (\mathbf{d} - \mathbf{Gm})^T \mathbf{W}_d (\mathbf{d} - \mathbf{Gm}) + (\mathbf{m} - \mathbf{m}_p)^T \mathbf{W}_M (\mathbf{m} - \mathbf{m}_p). \qquad (2.41)$$

For all the error functions (E_1 through E_5), formulas for least squares solutions can be easily derived. Menke (1984) gives expressions for most of these. Also, for more detailed discussions on methods of regularization, damped least squares, and the Backus–Gilbert approach, see Menke (1984) and Tarantola (1987).

2.4.4 Uncertainty estimates

Geophysical data almost always contain noise. In the analysis of linear inversion by the MLM, we assumed that the pdf is Gaussian and characterized by a data covariance matrix $\mathbf{C}_d$. The noise in the data will cause errors in the model parameter estimates. Also, the sensitivity of the error with respect to the model parameters decides the degree of uncertainty in model parameter estimates. This means that in some situations a range of model parameters may explain the data equally well even in the absence of any noise. A complete solution of the inverse problem can be described when in addition to estimating $\mathbf{m}_{est,}$ the uncertainty in the result is also quantified. Assuming that the pdf describing the answer (the model) is Gaussian with mean $\mathbf{m}_{est}$ and covariance cov($\mathbf{m}_{est}$) (this is also called the *posterior covariance* and will be described later in greater detail), we can do the following analysis: Recall that in the least squares method, we have

$$\mathbf{m}_{est} = \left[\mathbf{G}^T\mathbf{G}\right]^{-1}\mathbf{G}^T\mathbf{d} = \mathbf{G}^g\mathbf{d}. \tag{2.42}$$

Now, following Menke (1984, p. 59), we derive an expression for [cov $\mathbf{m}_{est}$]. From Eq. (2.42), we have

$$\left[\text{cov }\mathbf{m}_{est}\right] = \mathbf{G}^g\left[\text{cov }\mathbf{d}\right]\mathbf{G}^{gT} = \mathbf{G}^g\mathbf{C}_d\mathbf{G}^{gT}. \tag{2.43}$$

If we assume that the data are uncorrelated and that each datum has the same variance σ_d^2 we have

$$\mathbf{C}_d = \sigma_d^2\mathbf{I}. \tag{2.44}$$

Therefore, from Eqs. (2.43) and (2.44), we have

$$\left[\text{cov }\mathbf{m}_{est}\right] = \sigma_d^2\left[\mathbf{G}^T\mathbf{G}\right]^{-1}. \tag{2.45}$$

The second derivative of $E(\mathbf{m})$ with respect to $\mathbf{m}$ can be obtained from Eq. (2.28) as

$$\frac{\partial^2 E}{\partial \mathbf{m}^2} = 2\mathbf{G}^T\mathbf{G}. \tag{2.46}$$

From the preceding two equations, we obtain

$$\left[\text{cov }\mathbf{m}_{est}\right] = \sigma_d^2\left[\frac{1}{2}\frac{\partial^2 E}{\partial \mathbf{m}^2}\right]^{-1}_{\mathbf{m}=\mathbf{m}_{est}}. \tag{2.47}$$

Thus the covariance of model parameter estimates is given by the uncertainty in the data and the curvature of the error function with respect to the model parameters. The wider the error function near the minimum, the larger is the uncertainty in the model parameter estimate. Remember that the preceding expression for the covariance of model parameter estimates is based on the assumptions that the data are uncorrelated and that the prior pdf in data is Gaussian. Also, this analysis does not include any prior information on the model.

2.4.5 Regularization

In Section 2.4.3 we noted how we can impose constraints in the solution of an inverse problem. This essentially involves modification of the misfit term by adding a constraint term. In other words, a new error or misfit norm now consists of two terms, namely, a data norm E_d and a model norm E_m. The total error is given by a weighted sum of the two norms as follows:

$$E = E_d + \alpha E_m, \tag{2.48}$$

where α is a weighing term. This procedure is called *regularization*. Note that the choice of the term α, called *regularization weight*, is not trivial; we will discuss this later. The regularization weight controls the effect of model constraint on the solution of our inverse problem. All the procedures for imposing constraints described in the preceding section can be represented by Eq. (2.48).

Hansen (2007) provides an excellent description of regularization. He demonstrated with a simple example how an ill-posed linear inverse problem of estimating model $\mathbf{m}$ from data $\mathbf{d}$ can be stabilized by adding a model norm. Ill-posedness is caused by a large condition number of the generalized inverse matrix $\mathbf{G}^g$ that arises in a least squares problem [e.g., (Eq. 2.25)]. One approach to addressing this is to solve the least squares problem by adding the so-called side constraint that the solution norm must not exceed a certain value:

$$\min_{\mathbf{m}} \|\mathbf{Gm} - \mathbf{d}\|_2 \quad \text{subject to} \quad \|\mathbf{m}\|_2 \leq k. \tag{2.49}$$

Such a computed solution depends in a non-linear way on k, and with a proper choice of k, we can indeed compute a solution that is fairly close to the desired exact solution.

The objective of regularization theory is to provide efficient and numerically stable methods for including proper constraints that lead to useful stabilized solutions and to provide robust methods for choosing the optimal weight given to those constraints. The ultimate goal is to obtain a regularized solution that is a good approximation to the desired unknown solution (Hansen 2007).

In other words, for a linear system $\mathbf{d} = \mathbf{Gm}$, where the rank of $\mathbf{G}$ is smaller than the dimension of $\mathbf{m}$, some additional information is needed to find a satisfactory solution of $\mathbf{d} = \mathbf{Gm}$ because infinitely many solutions exist if there is one at all. In some applications, the acceptable solutions may require the "smoothness" of some function, curve, or surface constraints on $\mathbf{m}$. Tikhonov (1963) first modified the objective function as follows to impose the following constraint:

$$\|\mathbf{Gm} - \mathbf{d}\|_2 + \alpha \|\mathbf{m}\|_2 = \min \tag{2.50}$$

and showed that with a modified least squares problem such as the one shown in Eq. (2.50), the new solution is given by

$$\hat{\mathbf{m}} = (\mathbf{G}^T\mathbf{G} + \alpha\mathbf{I})^{-1}\mathbf{G}^T\mathbf{d}. \tag{2.51}$$

This formulation is also known as *damped least squares*, as described earlier in the text.

A general regularization of the form

$$E(\mathbf{m}) = \mathbf{e}^T\mathbf{e} = (\mathbf{d} - \mathbf{Gm})^T (\mathbf{d} - \mathbf{Gm}) + \mathbf{m}^T\mathbf{W}_m\mathbf{m}, \tag{2.52}$$

is known as *Tikhonov regularization*. When $\mathbf{W}_m = \mathbf{I}$, we have *zeroth-order* Tikhonov regularization, where we minimize an L_2 norm of model length together with data misfit. An appropriate form of $\mathbf{W}_m$ can be used in Eq. (2.52) to incorporate minimization of first or second derivative of the model, resulting in *first-* and *second-order* Tikhonov regularization. First, we define matrices $\mathbf{D}_1$ and $\mathbf{D}_2$ as follows:

$$\mathbf{D}_1 = \begin{bmatrix} -1 & 1 & \dots & & \\ & -1 & 1 & \dots & \\ & & -1 & 1 & \dots \\ & & & & \\ & & & & \\ & & & -1 & 1 \end{bmatrix}, \tag{2.53}$$

$$\mathbf{D}_2 = \begin{bmatrix} 1 & -2 & 1 & \dots & \\ & 1 & -2 & 1 & \dots \\ & & 1 & -2 & 1 & .. \\ & & & & \\ & & & 1 & -2 & 1 \end{bmatrix}, \tag{2.54}$$

for first and second-derivative operators, respectively. The $\mathbf{W}_m$ matrices corresponding to first- and second-order Tikhonov regularization are given by

$$\mathbf{W}_m = \mathbf{D}_1^T\mathbf{D}_1, \tag{2.55}$$

and

$$\mathbf{W}_m = \mathbf{D}_2^T\mathbf{D}_2, \tag{2.56}$$

respectively. Note that Eq. (2.52) is a special case of Eq. (2.41). The least squares solution of Eq. (2.41) or Eq. (2.52) can be obtained simply by setting the derivative of the objective function to zero, as described earlier.

The least squares solution for minimization of the error function given in Eq. (2.41) is

$$\mathbf{m}^{\text{est}} = \mathbf{m}_p + \left[\mathbf{G}^T \mathbf{W}_d \mathbf{G} + \alpha \mathbf{W}_m^{-1}\right] \mathbf{G}^T \mathbf{W}_d \left(\mathbf{d} - \mathbf{Gm}_p\right), \quad (2.57a)$$

and the solution for Eq. (2.52) is given by

$$\mathbf{m}^{\text{est}} = \left[\mathbf{G}^T \mathbf{G} + \alpha \mathbf{W}_m^{-1}\right] \mathbf{G}^T \mathbf{d}. \quad (2.57b)$$

Besides Tikhonov regularization, there are many other regularization methods with properties that make them better suited to certain problems or certain computers. One example is that of *maximum entropy regularization*, which is used in applications where a solution with positive elements is sought (e.g., Aster *et al.* 2005). One other very popular approach that can be related to Tikhonov regularization is based on the singular-value decomposition (SVD) described earlier in this chapter. Recall that the SVD of matrix $\mathbf{G}$ can be written as

$$\mathbf{G} = \mathbf{U}\Sigma\mathbf{V}^T, \quad (2.58)$$

where

$$\begin{aligned}
\mathbf{U} &= \left[u_1, u_2, ..., u_n\right] \\
\mathbf{V} &= \left[v_1, v_2, ..., v_n\right] \\
\mathbf{U}^T \mathbf{U} &= \mathbf{I}_n = \mathbf{U}^T \mathbf{V} \\
\Sigma &= \text{diag}(\sigma_1, \sigma_2, ..., \sigma_n)
\end{aligned}$$

The singular-value matrix has non-negative elements appearing in non-decreasing order such that

$$\sigma_1 \geq \sigma_2 \geq \sigma_3 \cdots \geq \sigma_n \geq 0.$$

Thus we have

$$\begin{aligned}
\mathbf{G} &= \mathbf{U}\Sigma\mathbf{V}^T, \\
\mathbf{G}^T &= \mathbf{V}\Sigma^T\mathbf{U}^T, \\
\mathbf{G}^T\mathbf{G} &= \mathbf{V}\Sigma^T\mathbf{U}^T\mathbf{U}\Sigma\mathbf{V}^T = \mathbf{V}\Sigma^T\Sigma\mathbf{V}^T, \\
\left(\mathbf{G}^T\mathbf{G}\right)^{-1} &= \left(\mathbf{V}^T\right)^{-1}\Sigma^{-1}\left(\Sigma^T\right)^{-1}\mathbf{V}^{-1}, \\
\left(\mathbf{G}^T\mathbf{G}\right)^{-1}\mathbf{G}^T &= \left(\mathbf{V}^T\right)^{-1}\Sigma^{-1}\left(\Sigma^T\right)^{-1}\mathbf{V}^{-1}\mathbf{V}\Sigma^T\mathbf{U}^T \\
&= \left(\mathbf{V}^T\right)^{-1}\Sigma^{-1}\mathbf{U}^T.
\end{aligned} \quad (2.59)$$

The estimated model is given as

$$\mathbf{m}^{\text{est}} = \left((\mathbf{V}^T)^{-1}\Sigma^{-1}\mathbf{U}^T\right)\mathbf{d} = \left(\mathbf{V}\Sigma^{-1}\mathbf{U}^T\right)\mathbf{d}. \quad (2.60)$$

If there are p non-zero singular values, we can write

$$\mathbf{U^T d} = \begin{bmatrix} \mathbf{u}_1^T \mathbf{d} \\ \mathbf{u}_2^T \mathbf{d} \\ \cdot \\ \cdot \\ \cdot \\ \mathbf{u}_p^T \mathbf{d} \end{bmatrix} ; \tag{2.61}$$

i.e., each element is a dot product of the first p columns of $\mathbf{U}$ with $\mathbf{d}$.

Thus we have

$$\boldsymbol{\Sigma}^{-1}\mathbf{U^T d} = \begin{bmatrix} \dfrac{1}{\sigma_1}\mathbf{u}_1^T \mathbf{d} \\ \dfrac{1}{\sigma_2}\mathbf{u}_2^T \mathbf{d} \\ \cdot \\ \cdot \\ \cdot \\ \dfrac{1}{\sigma_p}\mathbf{u}_p^T \mathbf{d} \end{bmatrix} , \tag{2.62}$$

and therefore,

$$\mathbf{m}^{\mathrm{est}} = \mathbf{V} \begin{bmatrix} \dfrac{1}{\sigma_1}\mathbf{u}_1^T \mathbf{d} \\ \dfrac{1}{\sigma_2}\mathbf{u}_2^T \mathbf{d} \\ \cdot \\ \cdot \\ \cdot \\ \dfrac{1}{\sigma_p}\mathbf{u}_p^T \mathbf{d} \end{bmatrix} . \tag{2.63}$$

The preceding equation clearly demonstrates why small singular values can have a large effect on the least squares solution in the presence of noise. Thus one straightforward approach to regularization is to eliminate small singular values of the system or add small values to each of the singular values. It turns out that this is exactly what is done in a damped least squares formulation that is equivalent to zeroth-order Tikhonov regularization.

Recall that the damped least squares solution is given by

$$(\mathbf{G^T G} + \alpha \mathbf{I})\mathbf{m} = \mathbf{G^T d}, \tag{2.64}$$

which can be written in terms of SVD as follows:

$$\left(\mathbf{V\Sigma^T U^T U\Sigma V^T} + \alpha \mathbf{I}\right)\mathbf{m} = \mathbf{V\Sigma^T U^T d},$$

or

$$\left(\mathbf{V\Sigma^T \Sigma V^T} + \alpha \mathbf{I}\right)\mathbf{m} = \mathbf{V\Sigma^T U^T d}. \tag{2.65}$$

Thus we have

$$m_\alpha = \sum_{i=1}^{b} \frac{\sigma_i^2}{\sigma_i^2 + \alpha} \frac{\mathbf{U_i^T d}}{\sigma_i} \mathbf{V}_i. \tag{2.66}$$

Note that the effect of the regularization weight is that in the denominator it gets added to the singular values. In other words, it adds a positive number to all the squares of the singular values. Thus the numerical instability due to small singular values is avoided.

2.4.5.1 *Method for choosing the regularization parameter*

We observe that the regularization parameter α controls the weight of the constraint or the *model norm* relative to the *residual* or *data norm*. Note that if too much regularization, or damping, is imposed on the solution, then it will not fit the given data $\mathbf{d}$ properly, and the residual $\|\mathbf{Gm} - \mathbf{d}\|_2$ will be too large. On the other hand, if too little regularization is imposed, then good fit will be found, but the solution will be dominated by the contributions from the data errors. The value of the parameter α should therefore be chosen carefully.

In most applications, several trial inversions are run with different values of regularization weight, and an optimal weight is chosen based on a qualitative evaluation of the quality of the inversion results. For linear problems, once a regularization weight is found to be satisfactory, it can be applied to estimate model updates.

The problem of estimation of regularization weights for linear problems has been investigated by several researchers and reported mainly in the mathematics and optimization literature. There exist various methods of choosing regularization weights; most of these methods are problem-specific and based on heuristics and possibly "surrounded by a sense of mystery" (Neumaier 1998). In essence, the methods of determining optimal regularizing weight can be broadly classified into two categories (Sen and Roy 2003):

- Methods that do not require prior knowledge of data error or noise:
     L-curve (Hansen and O'Leary 1993)
     Generalized cross-validation (Wahba 1990)
- Methods that do require prior knowledge of data error or noise:
     Discrepancy principle (Morozov 1986)
     Modified discrepancy principle (Engl 1987)

Some of these methods have been summarized in Hansen (1992).

The L-curve

Details of the L-curve technique are described in a series of papers by Hansen (e.g., Hansen 1992; Hansen and O'Leary 1993). It is interesting to note that if we plot with a suitable plotting scale the value of the model misfit versus, the value of the data misfit appearing on the right-hand side of Eq. (2.48) for various values of α, it generates an L-shaped curve. In other words, the L-curve technique is a graphic representation of model misfit and data misfit using a suitable plotting scale for varying values of the regularization weight. The very shape of the curve suggests its name. This is probably the most intuitive, simplest, and most attractive method of determining regularization weight. Further, the method is generic and can be used for determining multiple regularization weights for multiply constrained problems (Belge *et al.* 1998). The corner of the curve (point of maximum curvature) corresponds to the optimal value of the regularization weight (Figure 2.3). The common plotting scales are square root and logarithmic. However, it is observed that the shape of the L-curve is scale-dependent. It may degenerate its shapes in different scales of definition, and the characteristic point of the L-curve, the corner of it, may be obscured. In addition, Reginska (1996) showed that the L-curve defined in some scales lacks theoretical sufficiency to allow the corner of the curve to correspond to an optimal regularization weight. Once we decide on a scale of plotting, the L-curve technique comprises the following steps: We start with a set of α values; then for each α, we get the corresponding $\mathbf{m}^{(\alpha)}$. Then we compute the data misfit and the model misfit terms to generate an L-curve and look for an appropriate algorithm to locate the corner of the L-curve that corresponds to an optimal α.

Generalized cross-validation (GCV) method

The generalized cross-validation (GCV) method originally proposed by Craven and Wahba (1979) is a data-driven method in which a GCV function is defined as

$$V(\alpha) = \frac{\dfrac{1}{ND}\left\| \mathbf{d} - \mathbf{Gm}^{(\alpha)} \right\|_2^2}{\left[\dfrac{1}{ND} Tr(\mathbf{I} - \mathbf{A}(\alpha)) \right]^2}, \tag{2.67}$$

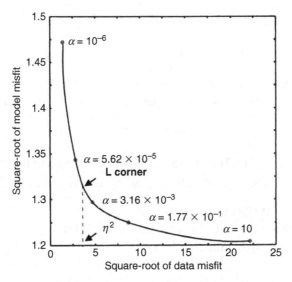

Figure 2.3 Plot of a typical L-curve, which is a plot of the model norm against the data norm. Note that points in the curve were generated for different values of the regularization weight. The optimal value of the regularization weight is picked from the corner, as shown in the figure.

where $\mathbf{A}(\alpha) = \mathbf{G}(\mathbf{G}^T\mathbf{G} + \alpha\mathbf{I})^{-1}\mathbf{G}^T$ is known as the *influence matrix*. The minimum of the GCV function $V(\alpha)$ would correspond to an optimal α. Note that for a large-scale problem, generation of the influence matrix and computation of its trace become highly intensive. However, Girard (1989) and subsequently Hutchinson (1989) proposed a Monte Carlo integration method in determining the trace of the matrix, which leads to a new technique known as *randomized GCV* (RGCV). The technique is based on the fact that for any symmetric matrix $\mathbf{B}$ of order NM and for vector $\mathbf{w}$ of dimension NM such that every entry of $\mathbf{w}$ is an independent sample from a random variable ξ with zero mean and unit variance, the following expression is valid:

$$E\left[\mathbf{w}^T\mathbf{B}\mathbf{w}\right] = Tr(\mathbf{B}), \tag{2.68}$$

where $E[\bullet]$ stands for expectation or mean of the function. Hence the randomized GCV function is defined as

$$V(\alpha) = \frac{\dfrac{1}{ND}\left\|\mathbf{d} - \mathbf{G}\mathbf{m}^{(\alpha)}\right\|_2^2}{\left[\dfrac{1}{ND}\left(\mathbf{w}^T\mathbf{w} - \mathbf{w}^T\mathbf{G}\theta_\alpha\right)\right]^2}, \tag{2.69}$$

where, the θ_α satisfies the following equation:

$$\mathbf{Gv}_\alpha = \mathbf{A}(\alpha)\mathbf{w}. \tag{2.70}$$

The main ingredient in RGCV computation is to determine θ_α by replacing $\mathbf{d}$, the right-hand term of the following equation with a random vector $\mathbf{w}$:

$$\mathbf{Gm}^{(\alpha)} = \mathbf{d}. \tag{2.71}$$

Equations (2.69) and (2.71) are the basis for computation of the RGCV function. Thus, following are the steps in optimal α selection using the RGCV technique:

- Select a series of α values.
- Compute for each α, $\mathbf{m}^{(\alpha)}$, θ_α, and RGCV function $V(\alpha)$.
- Find an optimal α by minimizing $V(\alpha)$.

Morozov's discrepancy principle

Morozov's discrepancy principle states that if η, the noise estimate in data, is known, then it is possible to determine an optimal regularization weight α through an iterative search such that in each iterative step k, a new value of α selected from a sequence $\{\alpha_o q\}$, with α_o and $q > 0$, will correspond to an optimizer $\mathbf{m}^{(\alpha)}$ that would satisfy the following discrepancy equation:

$$\left\| \mathbf{d}_\eta - \mathbf{Gm}^{(\alpha_k)} \right\|_2^2 \leq \zeta \eta^2, \tag{2.72}$$

where $\zeta > 1$ is any arbitrary constant. However, one needs to solve the linear problem many times to find an optimal α. Therefore, in a large-scale problem, such a method is prohibitively expensive.

Engl's modified discrepancy principle

This is a variant of the Morozov's discrepancy principle. This principle states that if η, the noise estimate in data, is known, then it is possible to have an a posteriori estimate of an optimal α that corresponds to the solution $\mathbf{m}^{(\alpha)}$ and satisfies following modified discrepancy equation:

$$\left\| \mathbf{G}^T\mathbf{Gm}^{(\alpha)} - \mathbf{G}^T\mathbf{d}_\eta \right\|_2^2 = \eta^a \alpha^{-b}, \tag{2.73}$$

provided that $\mathbf{G}^T\mathbf{d}_\eta \neq 0$, where a and b are arbitrary constants satisfying the constraint equation

$$\frac{3}{2}a - 2 = b \geq 1. \tag{2.74}$$

Thus, even for a large-scale problem, Engl's modified discrepancy principle is cost effective in estimating regularization weight. The main drawback of the method is the requirement for a noise estimate in the data, which is difficult to obtain in many geophysical inverse problems.

2.4.6 General L_p Norm

In our discussion so far, we used the L_2 norm to define our objective function; see, e.g., Eq. (2.57), in which both data and model misfits use the L_2 norm. The L_2 norm is associated with the normal distribution and is the most commonly used measure of error. However, in many applications, it may be more appropriate to use some other order norm. In a seismic deconvolution problem, for example, one can impose sparsity of spikes by using an L_1 norm of the reflectivity model parameters. In such problems, we need to use a mixed norm such that the data error is measured by an L_2 norm, whereas the minimum model length constraint is imposed by an L_1 norm. We recognize that the minimization of the L_1 norm is not trivial because the error function is no longer differentiable. For our linear inverse problem, we have

$$E = \|\mathbf{d} - \mathbf{Gm}\|_1.$$
(2.75)

Several algorithms have been developed to address this – these include iterative reweighted least squares (IRLS), total variation regularization, and the Huber norm.

2.4.6.1 IRLS

As discussed earlier, the first step in deriving a least squares formula is to take a derivative of E in Eq. (2.48) with respect to model vector $\mathbf{m}$ (e.g., Aster *et al.* 2005). We rewrite Eq. (2.75) as follows:

$$E = \|\mathbf{r}\|_1,$$
(2.76)

where $\mathbf{r}$ is the data residual vector. Note that the derivative of $\mathbf{r}$ with respect to $\mathbf{m}$ is discontinuous at a point where $\mathbf{r} = \mathbf{0}$. For now, let us ignore that fact, and we obtain the following derivative:

$$\frac{\partial E}{\partial \mathbf{m}} = \mathbf{G}^T \mathbf{Rr} = \mathbf{G}^T \mathbf{R}(\mathbf{d} - \mathbf{Gm}),$$
(2.77)

where

$$R_{ij} = \begin{cases} \dfrac{1}{|\mathbf{r}|} & \text{for } i = j, \\ 0 & \text{for } i \neq j, \end{cases}$$
(2.78)

is a diagonal matrix. Setting Eq. (2.77) equal to zero yields

$$\mathbf{G}^T\mathbf{RGm} = \mathbf{G}^T\mathbf{Rd}. \tag{2.79}$$

The preceding system of equations is very similar to the systems we have been discussing. There is, however, one very important difference. The elements of matrix $\mathbf{R}$ now depends on model $\mathbf{m}$. Thus it is a non-linear problem, and therefore, none of the developments reported in earlier sections are applicable. The ILRS method circumvents this issue by using a very simple trick. We first evaluate a solution using simple least squares with an L_2 norm. This solution is then used to compute the matrix $\mathbf{R}$ and is solved again using simple least squares (Eq. 2.77). This process approximates an L_2-norm solution and will fail when the residual becomes zero. In many applications, this has been found to be extremely useful.

2.4.6.2 Total variation regularization (TVR)

In a TVR approach (e.g., Rudin *et al.* 1992), the difficulty of differentiating a function of absolute value is avoided by approximating the function by smoothing as follows:

$$E(m) = |m| = \sqrt{m^2 + \beta}, \tag{2.80}$$

where β is a suitably chosen constant. The derivative of Eq. (2.80) can now be easily computed as follows:

$$\frac{\partial E}{\partial m} = \frac{m}{\sqrt{m^2 + \beta}}. \tag{2.81}$$

For our mixed-norm problem, we have

$$E(\mathbf{m}) = (\mathbf{d} - \mathbf{Gm})^T (\mathbf{d} - \mathbf{Gm}) + \alpha \mathbf{m}^T \mathbf{m} \approx (\mathbf{d} - \mathbf{Gm})^T (\mathbf{d} - \mathbf{Gm}) + \alpha \sqrt{\mathbf{m}^T \mathbf{m} + \beta}. \tag{2.82}$$

Taking the derivative of Eq. (2.82) and setting it equal to 0, we have

$$(\mathbf{G}^T\mathbf{G} + \alpha \mathbf{R})\mathbf{m} = \mathbf{G}^T\mathbf{d}, \tag{2.83}$$

where

$$\mathbf{R} = \frac{\mathbf{m}}{\sqrt{\mathbf{m}^T \mathbf{m} + \beta}}. \tag{2.84}$$

Thus we have a system of equations very similar to those described earlier.

Following the procedure described in this section, we can now derive a solution for a general L_p norm. Recall that the L_p norm objective function is given as

$$E_p = |\mathbf{Gm} - \mathbf{d}|^p = \left[\sum_{i-1}^{ND} |r_i|^p \right].$$ (2.85)

Note that

$$\frac{d}{dm} |m|^p = pm|m|^{p-2},$$ (2.86)

and therefore, we have

$$\nabla E_p = \mathbf{G}^T \mathbf{R} (\mathbf{Gm} - \mathbf{d}),$$ (2.87)

with

$$\mathbf{R} = \text{diag} \left\{ |r_i|^{p-2} \right\}.$$ (2.88)

Setting Eq. (2.83) equal to 0 yields

$$\mathbf{G}^T \mathbf{R}\, \mathbf{Gm} = \mathbf{G}^T \mathbf{Rd},$$ (2.89)

and this can again be solved by IRLS.

2.5 Iterative methods for non-linear problems: local optimization

The inverse problem for the system $\mathbf{d} = \mathcal{g}(\mathbf{m})$ can be reduced to a set of linear equations in cases where the forward problem is linear and in the case of weak non-linearity when perturbations in the data are linearly related to perturbations in the model parameters. We have seen in earlier sections that in such a case the solution of the inverse problem by the method of least squares or by the SVD method is a one-step procedure, and the model estimates and their uncertainties can be easily computed. Such linear methods do not work when the assumption of linearity is invalid.

For a non-linear problem, we write an L_2 objective function as follows:

$$E_d(\mathbf{m}) = (\mathbf{d} - \mathcal{g}(\mathbf{m}))^T \mathbf{W}_d (\mathbf{d} - \mathcal{g}(\mathbf{m})),$$ (2.90)

or, with model constraints, as

$$E(\mathbf{m}) = (\mathbf{d} - \mathcal{g}(\mathbf{m}))^T \mathbf{W}_d (\mathbf{d} - \mathcal{g}(\mathbf{m})) + \alpha (\mathbf{m} - \mathbf{m}_p)^T \mathbf{W}_M (\mathbf{m} - \mathbf{m}_p).$$ (2.91)

The solution of an inverse problem is then equivalent to finding a set of model parameters for which the objective function has the minimum value – a problem known as *optimization*.

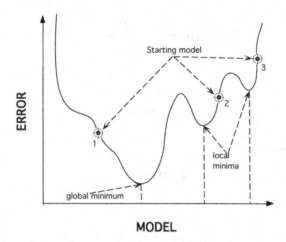

Figure 2.4 A hypothetical error function as a function of the model (the model is assumed to have only one model parameter) showing several minima. The iterative gradient method will find the global minimum only when starting at position 1. Others will end up in secondary minima of the error function.

The solution of an optimization problem is a set of allowed values of the variables (model parameters) for which the objective or the misfit function assumes an "optimal" value (Gill *et al.* 1981). Optimization usually involves maximizing or minimizing; for example, we may wish to maximize correlation or minimize data residual. Optimization methods are designed to search for the minimum of an objective function in an efficient manner. Obviously, such search algorithms depend on the shape or assumed shape of the objective function. As an example, we show a hypothetical error function in Figure 2.4 for a single model parameter. Note that the function is characterized by multiple peaks and troughs; this is very typical for non-linear problems. Each of the minima shown in the figure is called a *local minimum*, and the minimum of all the local minima is the *global minimum*. Once a suitably defined objective function is chosen [e.g., Eq. (2.90) or Eq. (2.91)], we would ideally like to find the global minimum of the objective function.

Several methods exist to search for the minima of a function of multiple variables. Referring to Figure 2.4, we note that it may be possible to reach the stationary point of the error surface that corresponds to the minimum error that may be approached by local linearization around a starting model by iteratively computing the model updates and matching the data. This is the case of *quasi-linearity*, and as demonstrated in the figure, success in finding the global minimum depends on the choice of the starting model. If the starting model is far from the global minimum, it is likely that the algorithm will get trapped into one of the local minima. These

Local Optimization – A model algorithm

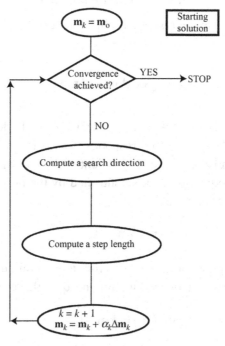

Figure 2.5 A model algorithm for local optimization.

methods are called *local optimization methods* and are the most popular optimization methods. Typically, these are deterministic methods that make use of local slope or curvature or both of the cost function to compute an update to the current model. These methods are calculus-based and always attempt to go in a downhill direction – this is why these are often called *greedy algorithms*.

Most local optimization methods can be described by the flowchart shown in Figure 2.5. Given a starting model, first it computes a search direction and a step length. A new model is generated and is evaluated; the process is repeated until there is no change in the model update or the cost function reaches a certain minimum value. At each iteration we check to make sure that the error is decreasing (hence the name *greedy*); if it increases, the algorithm stops. Local optimization methods vary depending on the way they compute the search direction. The most popular local optimization methods include Newton's method, steepest descent, and conjugate gradient, which will be described later in this section. Some local optimization methods assume that the shape of the objective function is quadratic, which results in great simplification of the algorithm. Let us briefly review the quadratic function first.

2.5.1 Quadratic function

Let us assume that we can represent our objective function as a quadratic function as follows:

$$E(\mathbf{m}) = \frac{1}{2}\mathbf{m}^T \mathbf{A}\mathbf{m} - \mathbf{b}^T \mathbf{m} + c, \qquad (2.92)$$

where $\mathbf{A}$ is a matrix, $\mathbf{b}$ is a vector, and c is a scalar. For a Taylor's series representation of the function, the matrix $\mathbf{A}$ corresponds to a matrix of second partial derivatives, and $\mathbf{b}$ corresponds to a vector of gradient of the objective function. Note that the shape of the objective function is controlled by the properties of matrix $\mathbf{A}$. If matrix $\mathbf{A}$ is positive definite, i.e.,

$$\mathbf{m}^T \mathbf{A}\mathbf{m} > 0, \qquad (2.93)$$

the cost function is a paraboloid with a well-defined minimum. When $\mathbf{A}$ is a symmetric matrix, the derivative of the function in Eq. (2.92) is given as

$$\nabla E(\mathbf{m}) = \mathbf{A}\mathbf{m} - \mathbf{b}. \qquad (2.94)$$

Setting this equation to 0 results in the following linear system of equations:

$$\mathbf{A}\mathbf{m} = \mathbf{b}. \qquad (2.95)$$

Thus, finding a minimum of a quadratic function is equivalent to solving a linear system.

2.5.2 Newton's method

The idea behind Newton's method is that the error function $E(\mathbf{m})$ being minimized is approximated locally by a quadratic function, and this approximating function is minimized exactly. It is assumed that the error $E(\mathbf{m}_{k+1})$ at a point $\mathbf{m}_{k+1}$ can be approximated by the following truncated Taylor's series:

$$E(\mathbf{m}_{k+1}) \cong E(\mathbf{m}_k) + (\mathbf{m}_{k+1} - \mathbf{m}_k)^T \nabla E(\mathbf{m}_k) + \frac{1}{2}(\mathbf{m}_{k+1} - \mathbf{m}_k)^T \nabla\nabla E(\mathbf{m}_k)(\mathbf{m}_{k+1} - \mathbf{m}_k).$$

$$(2.96)$$

The point $\mathbf{m}_{k+1}$ that minimizes this truncated series must satisfy

$$\nabla E(\mathbf{m}_k) + \nabla\nabla E(\mathbf{m}_k)(\mathbf{m}_{k+1} - \mathbf{m}_k) = 0. \qquad (2.97)$$

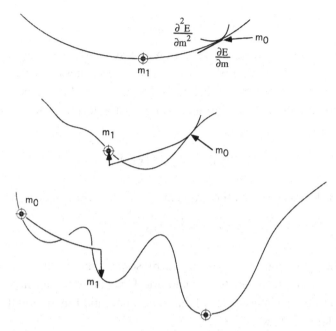

Figure 2.6 Three 1D objective functions of varying complexity that are to be min-imized. Random methods simply evaluate the curve at many points, the smallest of which is hoped to be the minimum. Gradient methods use the slope to find the nearest minimum, and Newton's method uses the slope (first derivative) and curva-ture (second derivative) to make a parabolic fit. For increasingly non-parabolic functions (middle), Newton's method may require more than one iteration to con-verge, and for extremely non-parabolic functions (bottom), it will fail unless the process is repeated from many different starting points. (From Wood 1993.)

This gives

$$\mathbf{m}_{k+1} = \mathbf{m}_k - \left[\nabla\nabla E(\mathbf{m}_k)\right]^{-1} \nabla E(\mathbf{m}_k) = \mathbf{m}_k - \mathbf{H}^{-1}(\mathbf{m}_k)\nabla E(\mathbf{m}_k), \quad (2.98)$$

where $\mathbf{H}$ is the matrix of second derivatives called the *Hessian*. Referring to our generic algorithm described in Figure 2.6, the search direction of Newton's method is computed by a product of the inverse Hessian and the gradient vector. Where the error function is truly quadratic, Newton's method converges in only one iteration (Figure 2.6). For practical applications, the model update is multiplied by a suit-able step length α_k:

$$\mathbf{m}_{k+1} = \mathbf{m}_k - \left[\nabla\nabla E(\mathbf{m}_k)\right]^{-1} \nabla E(\mathbf{m}_k) = \mathbf{m}_k - \alpha_k \mathbf{H}^{-1}(\mathbf{m}_k)\nabla E(\mathbf{m}_k). \quad (2.99)$$

The step length can be determined using a line search or a bisection method (Press *et al.* 1989). Note that numerical computation of gradient and especially the Hessian is not a trivial task.

2.5.3 Steepest descent

In the method of steepest descent, we start with a starting model, say, $\mathbf{m}_0$, and take a series of steps until the algorithm converges. At each iteration, we move along the direction in which the error changes most rapidly, which happens to be the direction of the gradient. Thus the steepest-descent algorithm can be given by the following equation:

$$\mathbf{m}_{k+1} = \mathbf{m}_k + \alpha_k \nabla E(\mathbf{m}_k). \tag{2.100}$$

For simplicity, let us represent the gradient of the objective function as follows:

$$E'(\mathbf{m}_k) = \nabla E(\mathbf{m}_k). \tag{2.101}$$

In general, the step length must be chosen by some standard method such as line search or bisection. However, for a quadratic function, one can find an analytic expression for the step length. Recall that for a quadratic function with a symmetric, positive-definite $\mathbf{A}$, we have

$$E'(\mathbf{m}_k) = \mathbf{A}\mathbf{m}_k - \mathbf{b} = -(\mathbf{b} - \mathbf{A}\mathbf{m}_k) = -\mathbf{r}_k. \tag{2.102}$$

The appropriate value of step length can be found by taking a derivative of the error function with respect to the step length and setting it to 0. Thus we have

$$\frac{d}{d\alpha_k} E(\mathbf{m}_k) = E'(\mathbf{m}_k)^T \frac{d}{d\alpha_k} \mathbf{m}_k = E'(\mathbf{m}_k)^T \mathbf{r}_{k-1} = 0. \tag{2.103}$$

Thus we notice that the value of α should be so chosen that the search direction at iteration $k+1$ is orthogonal to the previous search direction k. Thus we have

$$\begin{aligned} \mathbf{r}_{k+1}{}^T \mathbf{r}_k &= 0, \\ \left(\mathbf{b} - \mathbf{A}\mathbf{m}_{k+1}\right)^T \mathbf{r}_k &= 0, \\ \left(\mathbf{b} - \mathbf{A}\left(\mathbf{m}_k + \alpha_k \mathbf{r}_k\right)\right)^T \mathbf{r}_k &= 0, \end{aligned} \tag{2.104}$$

which gives

$$\alpha_k = \frac{\mathbf{r}_k^T \mathbf{r}_k}{\mathbf{r}_k^T \mathbf{A}\mathbf{r}_k}. \tag{2.105}$$

Thus we have

$$\mathbf{m}_{k+1} = \mathbf{m}_k + \frac{\mathbf{r}_k^T \mathbf{r}_k}{\mathbf{r}_k^T \mathbf{A}\mathbf{r}_k} \mathbf{r}_k. \tag{2.106}$$

The steepest-descent method zigzags its way to the optimal position such that each successive direction is perpendicular to the previous direction. One serious difficulty encountered by this approach is that for error functions with elongated valleys, the algorithm takes very small steps and therefore takes a long time to converge. The conjugate gradient method, which we describe next, circumvents this problem.

2.5.4 Conjugate gradient

As with the steepest-descent approach, let us assume that the function to be minimized is a quadratic given by Eq. (2.92), with the gradient given by Eq. (2.94). We write an updated formula at the first iteration as follows:

$$\mathbf{m}_1 = \mathbf{m}_0 + \alpha_0 \mathbf{p}_0, \qquad (2.107)$$

where $\mathbf{p}_0$ is the update direction called *conjugate direction*. At the first iteration we set $\mathbf{p}_0 = \mathbf{r}_0$ such that

$$\alpha_0 = \frac{\mathbf{r}_0^T \mathbf{p}_0}{\mathbf{p}_0^T A \mathbf{p}_0},$$

$$\mathbf{m}_1 = \mathbf{m}_0 + \frac{\mathbf{r}_0^T \mathbf{p}_0}{\mathbf{p}_0^T A \mathbf{p}_0} \mathbf{p}_0. \qquad (2.108)$$

Note that $\mathbf{r}_0^T \mathbf{p}_0$ is the projection of residual $\mathbf{r}_0$ along the direction $\mathbf{p}_0$. Now we need to find a new direction

$$\mathbf{p}_1 = \mathbf{r}_1 + \beta_1 \mathbf{p}_0, \qquad (2.109)$$

where β_1 is a new step length along the conjugate direction. In addition, we impose the following conjugacy condition (Press *et al.* 1989):

$$\mathbf{p}_1^T A \, \mathbf{p}_0 = 0; \qquad \mathbf{p}_0^T A \, \mathbf{p}_1 = 0. \qquad (2.110)$$

Thus we have

$$\mathbf{p}_0^T A \mathbf{p}_1 = \mathbf{p}_0^T A \left(\mathbf{r}_1 + \beta_1 \mathbf{p}_0 \right) = 0;$$

$$\beta_1 = -\frac{\mathbf{p}_0^T A \mathbf{r}_1}{\mathbf{p}_0^T A \mathbf{p}_0} = -\frac{\mathbf{r}_1^T A \mathbf{p}_0}{\mathbf{p}_0^T A \mathbf{p}_0}. \qquad (2.111)$$

By induction, we can write

$$\beta_k = -\frac{\mathbf{r}_k^T A \mathbf{p}_{k-1}}{\mathbf{p}_{k-1}^T A \mathbf{p}_{k-1}}. \qquad (2.112)$$

The algorithm can now be summarized by the following steps:

$$\mathbf{m}_{k+1} = \mathbf{m}_k + \alpha_k \mathbf{p}_k, \qquad \text{model update}$$

$$\alpha_k = -\frac{\mathbf{r}_k^T \mathbf{A} \mathbf{p}}{\mathbf{p}_k^T \mathbf{A} \mathbf{p}_k}, \qquad \text{step length}$$

$$\mathbf{p}_k = \mathbf{r}_k + \beta_k \mathbf{p}_{k-1}, \qquad \text{new direction} \qquad (2.113)$$

$$\beta_k = -\frac{\mathbf{r}_k^T \mathbf{A} \mathbf{p}_{k-1}}{\mathbf{p}_{k-1}^T \mathbf{A} \mathbf{p}_{k-1}}, \qquad \text{step length for new direction}$$

$$\mathbf{r}_{k+1} = \mathbf{r}_k - \alpha_k \mathbf{A} \mathbf{p}_k. \qquad \text{update the residual}$$

2.5.5 Gauss–Newton

Having discussed some of the most popular local optimization methods, such as Newton's method, steepest descent, and conjugate gradient, let us revisit our objective function in Eq. (2.91), for which we have

$$\mathbf{d}^{\text{syn}} = \mathscr{g}(\mathbf{m}),$$

$$\mathbf{G}_n = \left. \frac{\partial(\mathscr{g}(\mathbf{m}))}{\partial \mathbf{m}} \right|_{m=m_n}, \qquad (2.114)$$

$$G_n^{ij} = \left. \frac{\partial(\mathscr{g}^i)}{\partial m^j} \right|_{m=m_n},$$

where $\mathbf{G}_n$ is a matrix of partial derivatives of the synthetic data with respect to each model parameter evaluated at model $\mathbf{m} = \mathbf{m}_n$. We will closely follow Tarantola (1987) to derive a surprising result. The derivative of the cost function is given by

$$\nabla E|_{m=m_n} = \mathbf{G}_n^T \mathbf{W}_d \left(\mathscr{g}(\mathbf{m}_n) - \mathbf{d} \right) + \mathbf{W}_m (\mathbf{m}_n - \mathbf{m}_p), \qquad (2.115)$$

and the second derivative is given as

$$H|_{m=m_n} = \mathbf{G}_n^T \mathbf{W}_d \mathbf{G}_n + \mathbf{W}_m + \nabla \mathbf{G}_n^T \mathbf{W}_d \left(\mathscr{g}(\mathbf{m}_n) - \mathbf{d} \right). \qquad (2.116)$$

We ignore the third term based on the argument that the data residual is small, and therefore, we have

$$H|_{m=m_n} \approx \mathbf{G}_n^T \mathbf{W}_d \mathbf{G}_n + \mathbf{W}_m. \qquad (2.117)$$

Now we substitute Eqs. (2.115) and (2.116) into Eq. (2.99) to obtain the following updated formula:

$$\mathbf{m}_{k+1} = \mathbf{m}_k - \alpha_k \left(\mathbf{G}_n^T \mathbf{W}_d \mathbf{G}_n + \mathbf{W}_m \right)^{-1} \left[\mathbf{G}_n^T \mathbf{W}_d \left(\mathscr{g}(\mathbf{m}_n) - \mathbf{d} \right) + \mathbf{W}_m \left(\mathbf{m}_n - \mathbf{m}_p \right) \right].$$

(2.118)

This equation is Newton's formula for our general objective function with an approximate Hessian. Let us now examine some special cases. With

$$\mathbf{W}_m = 0 \quad \text{and} \quad \mathbf{W}_d = \mathbf{I},$$

we have

$$\mathbf{m}_{k+1} = \mathbf{m}_k - \alpha_k \left(\mathbf{G}_n^T \mathbf{G}_n \right)^{-1} \mathbf{G}_n^T \left(\mathscr{g}(\mathbf{m}_n) - \mathbf{d} \right), \quad (2.119)$$

where the model update part can be identified as

$$\Delta \mathbf{m} = \left(\mathbf{G}_n^T \mathbf{G}_n \right)^{-1} \mathbf{G}_n^T \Delta \mathbf{d}. \quad (2.120)$$

This formula is the same as that for the least squares solution of a linear system derived earlier. The expression in Eq. (2.120) is called the *Gauss–Newton update formula*.

Geophysical inversion is a two-step procedure. In the first step, different statistical and physical theories are combined to define the function that needs to be minimized. The second step involves optimization, which is the mathematical technique used to determine the minimum of the specified function. In most of our discussions so far, we have considered physical theories to define a simple error function E that was then minimized using a linear method. However, due to uncertainties in the data, the physical theory, and prior models, both the data and the model can best be described as random variables, and the inversion can be described in the framework of probability theory. We briefly mentioned this approach in the sections on maximum likelihood and prior information in the context of the linear inverse method. Franklin (1970) and Jackson (1979) incorporated these concepts into linear inverse theory and called it the *stochastic inversion approach*.

The classical approach to geophysical inversion involves assuming a linear relationship between the data and the model, and several useful concepts can be illustrated based on the principles of linear algebra. The concept of non-uniqueness can be very clearly illustrated by examining the length of the data and model vectors to decide whether a problem is underdetermined or overdetermined. Concepts of data and model resolution can also be developed, and the model resolution essentially describes the uncertainty in the derived result [see Menke (1984) for details]. Later, such a classical approach was modified into the MLM, in which the probabilistic model of error in the data was included, and then attempts were also made to use prior information on the model in describing the solution of an inverse problem.

Tarantola and Vallette (1982a, 1982b) and Tarantola (1987) formulated the inverse problem on a much stronger theoretical and statistical basis and showed how different states of information can be combined to define an answer to the inverse problem. Tarantola (1987) also showed how gradient-based optimization methods can be used in the search for the minimum of an error function. One other formulation that is also used in geophysical inverse problems and is very similar to Tarantola's formulation is the Bayesian formulation. Tarantola's formulation can also be derived from Bayes' rule (Duijndam 1988a, 1988b). A detailed description of the Bayesian inference models can be found in the text by Box and Tiao (1973), and a very readable version of the same for geophysical applications is given in Jackson and Matsura (1985) and Cary and Chapman (1988). We discussed Bayes' rule in Chapter 1 and showed how it can be used to update our knowledge when new data become available. Here we will closely follow Duijndam (1987) to describe the formulation of the inverse problem using Bayes' rule.

As usual, we will represent the model by a vector $\mathbf{m}$ and data by a vector $\mathbf{d}$ given by

$$\mathbf{m} = \left[m_1, m_2, m_3, \ldots, m_{NM} \right]^T$$

$$\mathbf{d} = \left[d_1, d_2, d_3, \ldots, d_{ND} \right]^T$$

consisting of elements m_i and d_i, respectively, each one of which is considered to be a random variable. The quantities NM and ND are the numbers of model parameters and data points, respectively, and the superscript T represents a matrix transpose. Following Tarantola's (1987) notation, we assume that $p(\mathbf{d}|\mathbf{m})$ is the conditional pdf of $\mathbf{d}$ for a given $\mathbf{m}$, $\sigma(\mathbf{m}|\mathbf{d})$ is the conditional pdf of $\mathbf{m}$ for a given $\mathbf{d}$, $p(\mathbf{d})$ is the pdf of the data $\mathbf{d}$, and $p(\mathbf{m})$ is the pdf of model $\mathbf{m}$ independent of $\mathbf{d}$. Then we have, from the definition of conditional probabilities,

$$\sigma(\mathbf{m}|\mathbf{d}) p(\mathbf{d}) = p(\mathbf{d}|\mathbf{m}) p(\mathbf{m}). \tag{2.121}$$

From the preceding equation we obtain the following equation for the conditional distribution of the model $\mathbf{m}$ given the data $\mathbf{d}$:

$$\sigma(\mathbf{m}|\mathbf{d}) = \frac{p(\mathbf{d}|\mathbf{m}) p(\mathbf{m})}{p(\mathbf{d})}, \tag{2.122}$$

which is the state of information on model $\mathbf{m}$ given the data $\mathbf{d}$. This equation is the so-called Bayes' rule. In geophysical inversion, the denominator term $p(\mathbf{d})$, which is independent of model $\mathbf{m}$, is a constant. Replacing the denominator with a constant in the preceding equation gives

$$\sigma(\mathbf{m}|\mathbf{d}) \propto p(\mathbf{d}|\mathbf{m}) p(\mathbf{m}). \tag{2.123}$$

When the measured data $\mathbf{d}_{\text{obs}}$ are substituted into the expression for the conditional pdf, $p(\mathbf{d}|\mathbf{m})$, it is called the *likelihood function* and denoted by $l(\mathbf{d}_{\text{obs}}|\mathbf{m})$ or simply $l(\mathbf{m})$, a function of $\mathbf{m}$. Thus the preceding equation can be written as (Box and Tiao 1973; Cary and Chapman 1988):

$$\sigma(\mathbf{m}|\mathbf{d}) \propto l(\mathbf{d}_{\text{obs}}|\mathbf{m}) p(\mathbf{m}). \tag{2.124}$$

The pdf $p(\mathbf{m})$ is the probability of the model $\mathbf{m}$ independent of the data; i.e., it describes the information we have on the model without knowledge of the data and is called the *prior distribution*. Similarly, the pdf $\sigma(\mathbf{m}|\mathbf{d})$ is a description of the model $\mathbf{m}$ for the given data and is called the *posterior probability density function* (PPD) when normalized. Clearly, the PPD is obtained by a product of the likelihood function and the prior distribution. Thus the prior knowledge in the model is modified by the likelihood function, which is determined by the observed data. Box and Tiao (1973) and Duijndam (1987) show different theoretical examples of how the posterior is influenced depending on the relative importance between the prior and the likelihood function. For a uniform prior, the PPD is determined primarily by the likelihood function. Only in very rare circumstances can we have a situation where the prior will dominate the likelihood function.

The choice of the likelihood function depends on the distribution of the noise or error in the data. Thus it requires prior knowledge of the error distribution of the data. Most inverse problems can be treated using a standard reduced model (Bard 1974; Duijndam 1987) given by the following equation:

$$\mathbf{d} = \mathbf{g}(\mathbf{m}) + \mathbf{n}, \tag{2.125}$$

where g is the forward modeling operator and the vector $\mathbf{n}$ contains error or noise samples in the data. We now assume that the noise $\mathbf{n}$ is independent of $g(\mathbf{m})$ (i.e., the forward model does not account for noise), characterized by the pdf $p(\mathbf{n})$. Clearly,

$$p(\mathbf{n}) = p(\mathbf{d} - \mathbf{g}(\mathbf{m})). \tag{2.126}$$

It is also obvious that the right-hand side of this equation corresponds to the conditional pdf $p(\mathbf{d}|\mathbf{m})$; i.e.,

$$p(\mathbf{d}|\mathbf{m}) = p(\mathbf{n}) = p(\mathbf{d} - g(\mathbf{m})). \tag{2.127}$$

The likelihood function $l(\mathbf{m})$ is given by the following equation:

$$l(\mathbf{m}) = p(\mathbf{n}) = p(\mathbf{d} = \mathbf{d}_{obs}|\mathbf{m}). \tag{2.128}$$

In geophysical inversion, a distinction is usually made between theoretical and observational errors (e.g., Tarantola 1987). While the observational errors can be attributed to instrumental errors, the theoretical error is caused by the use of an inexact theory in the prediction of theoretical data. For example, the use of 1D models could cause theoretical errors because in most cases the earth is truly three-dimensional.

Let us assume that the operator g_{th} denotes an exact forward modeling operator. In such a situation, the theoretical error vector $\mathbf{n}_{th}$ is given by

$$\mathbf{n}_{th} = g_{th}(\mathbf{m}) - g(\mathbf{m}). \tag{2.129}$$

Thus, from Eqs. (2.124) and (2.128), we obtain

$$\mathbf{d} = g_{th}(\mathbf{m}) - \mathbf{n}_{th} + \mathbf{n} \tag{2.130}$$

or

$$\mathbf{n}_{th} - \mathbf{n} = \mathbf{n}_o = \mathbf{d} - g_{th}(\mathbf{m}), \tag{2.131}$$

where $\mathbf{n}_o$ corresponds to the observational error, and the total error $\mathbf{n}$ is the sum of theoretical and observational errors; i.e.,

$$\mathbf{n} = \mathbf{n}_{th} = \mathbf{n}_o. \tag{2.132}$$

Here we emphasize that the distinction between the theoretical and observational errors is highly arbitrary. Any unmodeled feature in the data is termed *noise*. It is justified to say that an exact theory should be able to predict every phenomenon, including the malfunctioning of an instrument. Thus there is no need of a separate *observational error* term, and the entire noise in the data can be explained as theoretical error. On the other hand, it can also be argued that no theory is exact and that any phenomenon not predictable by the theory in use is observational error. Thus there is no need for a separate theoretical error. The question of whether such a distinction is important or necessary has not been investigated in the geophysical literature.

However, once we have decided on defining some error to be theoretical and the other to be observational, we can rewrite the conditional distribution (or the likelihood function) to show the effect of the two (Duijndam 1987). Thus, if we write

$$\mathbf{d}_{th} = g_{th}(\mathbf{m}), \tag{2.133}$$

where $\mathbf{d}_{th}$ are the data generated by an exact theory, we can use the definition of conditional and joint pdf (Chapter 1) to write the following equation:

$$p(\mathbf{d}|\mathbf{m}) = \int p(\mathbf{d}|\mathbf{d}_{th}, \mathbf{m}) p(\mathbf{d}_{th}|\mathbf{m}) dd_{th}. \tag{2.134}$$

Recall that $p(\mathbf{d}|\mathbf{d}_{th}, \mathbf{m})$ corresponds to the pdf of the observational error and may be considered to be independent of $\mathbf{m}$; i.e.,

$$p(\mathbf{d}|\mathbf{d}_{th}, \mathbf{m}) = p(\mathbf{d}|\mathbf{d}_{th}) = p(n_o) = p(\mathbf{d} - \mathbf{d}_{th}). \tag{2.135}$$

Also, $p(\mathbf{d}_{th}|\mathbf{m})$ corresponds to the pdf of noise due to theory errors; i.e.,

$$p(\mathbf{d}_{th}|\mathbf{m}) = p(\mathbf{n}_{th}) = p(\mathbf{d}_{th} - g(\mathbf{m})). \tag{2.136}$$

Thus we have

$$p(\mathbf{d}|\mathbf{m}) = \int p(\mathbf{d} - \mathbf{d}_{th}) p(\mathbf{d}_{th} - g(\mathbf{m})) dd_{th}, \tag{2.137}$$

or

$$p(\mathbf{d}|\mathbf{m}) = p(\mathbf{d} - \mathbf{d}_{th}) * p(\mathbf{d}_{th} - g(\mathbf{m})). \tag{2.138}$$

where * represents the convolution operator. Comparing with Eq. (2.128), we find the well-known property that the pdf of the sum of two independent variables is the convolution of the pdfs of the two variables.

Assuming Gaussian error, both in theory and in observations characterized by the covariances $\mathbf{C}_T$ and $\mathbf{C}_d$, respectively, the likelihood function takes the following form:

$$l[\mathbf{d}_{obs}|\mathbf{m}] \propto \exp(E(\mathbf{m})), \tag{2.139}$$

where $E(\mathbf{m})$ is the error function given by

$$E(\mathbf{m}) = \left[-\frac{1}{2}(\mathbf{d}_{obs} - g(\mathbf{m}))^T \mathbf{C}_D^{-1} (\mathbf{d}_{obs} - g(\mathbf{m})) \right], \tag{2.140}$$

where g is the forward modeling operator and $\mathbf{C}_D$ is called the *data covariance matrix*. The data covariance matrix consists of two parts, namely, the experimental uncertainty $\mathbf{C}_d$ and the modelization uncertainty or error due to theory $\mathbf{C}_T$ (Tarantola 1987); i.e.,

$$\mathbf{C}_D = \mathbf{C}_d + \mathbf{C}_T. \tag{2.141}$$

Equation (2.139) can be derived by substituting Gaussian pdfs for the two terms in Eq. (2.138).

Using Eqs. (2.124) and (2.125), the expression for the PPD can thus be written as

$$\sigma\left(\mathbf{m}|\mathbf{d}_{obs}\right) \propto \exp\left(-E(\mathbf{m})\right)p(\mathbf{m}). \tag{2.142}$$

or

$$\sigma\left(\mathbf{m}|\mathbf{d}_{obs}\right) = \frac{\exp\left(-E(\mathbf{m})\right)p(\mathbf{m})}{\int d\mathbf{m}\,\exp\left(-E(\mathbf{m})\right)p(\mathbf{m})}, \tag{2.143}$$

where the domain of integration covers the entire model space.

By substituting a Gaussian pdf for the prior $p(\mathbf{m})$ characterized by a prior model $\mathbf{m}_p$ and prior model covariance matrix $\mathbf{C}_m$

$$p(\mathbf{m}) \propto \exp\left[-\frac{1}{2}\left(\mathbf{m}-\mathbf{m}_p\right)^T \mathbf{C}_m^{-1}\left(\mathbf{m}-\mathbf{m}_p\right)\right] \tag{2.144}$$

into Eq. (2.142), and using Eqs. (2.139) and (2.140), we obtain the following expression for the PPD:

$$\sigma\left(\mathbf{m}|\mathbf{d}_{obs}\right) \propto \exp\left(E(\mathbf{m})\right), \tag{2.145}$$

where

$$E(\mathbf{m}) = -\frac{1}{2}\left[\left(\mathbf{d}-\mathbf{g}(m)\right)^T \mathbf{C}_D^{-1}\left(\mathbf{d}-\mathbf{g}(\mathbf{m})\right)+\left(\mathbf{m}-\mathbf{m}_p\right)^T \mathbf{C}_m^{-1}\left(\mathbf{m}-\mathbf{m}_p\right)\right]. \tag{2.146}$$

Once the PPD has been identified, as given by Eq. (2.143), the answer to the inverse problem is given by the PPD. This is merely a description of the problem. In reality, we are faced with the problem of estimating the PPD in a large multidimensional model space. The most accurate way of evaluating the PPD is to compute the right-hand side of the preceding equation at each point in model space. That is, evaluate the forward problem at each point in model space. This is, in general, a very time-consuming task unless the prior information helps us to restrict the model space to a small region. Note also that even with the simplifying assumptions of Gaussian pdfs for the data error and model prior, the resulting PPD (Eq. 2.145) is non-Gaussian. This is due to the presence of the term $g(\mathbf{m})$ in the likelihood function in Eq. (2.139) (Tarantola 1987).

One simplistic approach is to use an optimization technique to locate the minimum of the error function, which usually corresponds to the maximum of the PPD.

Such a method is called the *maximum a posteriori* (MAP) *estimation method*. If we assume that the PPD can be approximated by a Gaussian around its peak (assuming that there is one such peak), also called the *MAP point*, then the MAP point will also correspond to the mean of the Gaussian PPD. The posterior covariance matrix can be computed from the curvature of the error at the MAP point, and thus we have a complete description of the PPD. In general, however, the PPD may be multimodal, and such an approach will not work.

Even if the PPD were known, there is no way to display it in a multidimensional space. Therefore, several measures of dispersion and marginal density functions are often used to describe the answer. The marginal PPD of a particular model parameter, the posterior mean model, and the posterior model covariance matrix are given by

$$\sigma\left(m_i | \mathbf{d}_{obs}\right) = \int dm_1 \int dm_2 \int dm_2 \cdots \int dm_{i-1} \int dm_{i+1} \cdots \int dm_{NM} \sigma\left(\mathbf{m} | \mathbf{d}_{obs}\right), \quad (2.147)$$

$$\langle \mathbf{m} \rangle = \int d\mathbf{m} \ \mathbf{m} \ \sigma\left(\mathbf{m} | \mathbf{d}_{obs}\right), \quad (2.148)$$

and

$$\mathbf{C}_M' = \int d\mathbf{m} \left(\mathbf{m} - \langle \mathbf{m} \rangle\right)\left(\mathbf{m} - \langle \mathbf{m} \rangle\right)^T \sigma\left(\mathbf{m} | \mathbf{d}_{obs}\right). \quad (2.149)$$

All these integrals belong to the following general form:

$$I = \int d\mathbf{m} \ f\left(\mathbf{m}\right)\sigma\left(\mathbf{m} | \mathbf{d}_{obs}\right). \quad (2.150)$$

Thus estimation of marginal PPDs and mean and posterior covariance requires evaluation of multidimensional integrals of the type given in Eq. (2.150). We will describe several numerical methods of evaluating these in Chapter 8.

2.6 Solution using probabilistic formulation

Different forms of the PPD can be obtained based on the assumptions made on the form of the prior pdfs, the most common of which is the Gaussian, as given by Eqs. (2.145) and (2.146). Equation (2.145) describes the posterior PPD in model parameters, assuming that the pdfs describing prior information are Gaussian. Note that even at this stage, no assumption has been made about the linearity of the problem. However, following Tarantola (1987), Eq. (2.145) can be examined for different degrees of linearity, which we discuss next.

2.6.1 Linear case

When the forward problem is linear, we can replace the forward modeling operator g with the matrix $\mathbf{G}$ such that

$$\mathbf{d} = \mathbf{G}\,\mathbf{m}. \tag{2.151}$$

From Eqs. (2.145) and (2.146), we obtain

$$\sigma\left(\mathbf{m}|\mathbf{d}_{obs}\right) \propto \exp\left[-\frac{1}{2}\left[\left(\mathbf{Gm} - \mathbf{d}_{obs}\right)^T \mathbf{C}_D^{-1}\left(\mathbf{Gm} - \mathbf{d}_{obs}\right)\right.\right.$$
$$\left.\left. +\left(\mathbf{m} - \mathbf{m}_p\right)^T \mathbf{C}_m^{-1}\left(\mathbf{m} - \mathbf{m}_p\right)\right]\right]. \tag{2.152}$$

Following Tarantola (1987), we define

$$\langle\mathbf{m}\rangle = \left[\mathbf{G}^T\mathbf{C}_D^{-1}\mathbf{G} + \mathbf{C}_m^{-1}\right]^{-1}\left[\mathbf{G}^T\mathbf{C}_D^{-1}\mathbf{d}_{obs} + \mathbf{C}_m^{-1}\,\mathbf{m}_p\right]. \tag{2.153}$$

and

$$\mathbf{C}_{M'} = \left(\mathbf{G}^T\mathbf{C}_D^{-1}\mathbf{G} + \mathbf{C}_m^{-1}\right). \tag{2.154}$$

This gives

$$\sigma\left(\mathbf{m}|\mathbf{d}_{obs}\right) \propto \exp\left[-\frac{1}{2}\left(\mathbf{m} - \langle\mathbf{m}\rangle\right)^T \mathbf{C}_{M'}^{-1}\left(\mathbf{m} - \langle\mathbf{m}\rangle\right)\right]. \tag{2.155}$$

This shows that when the forward problem is linear, the PPD in the model space is Gaussian with mean <m> and posterior covariance matrix $\mathbf{C}'_M$ For a Gaussian PPD, the mean and covariance completely describe the distribution. Given the prior model $\mathbf{m}_p$, the prior covariance matrices $\mathbf{C}_m$ and $\mathbf{C}_D$, and the linear forward modeling operator $\mathbf{G}$, it is a one-step procedure to derive the answer to the inverse problem, i.e., to describe the posterior pdf in model space. It can be shown from Eq. (2.154) that

$$\langle\mathbf{m}\rangle = \mathbf{m}_p + \left[\mathbf{G}^T\mathbf{C}_D^{-1}\mathbf{G} + \mathbf{C}_m^{-1}\right]^{-1}\mathbf{G}^T\mathbf{C}_D^{-1}\left(\mathbf{d}_{obs} - \mathbf{d}_p\right), \tag{2.156}$$

where $\mathbf{d}_p = \mathbf{Gm}_p$.

When we set $\mathbf{C}_m^{-1}$ to 0 and $\mathbf{C}_D^{-1}$ to an identity matrix, we get back the standard least squares formula for model estimates $\mathbf{m}_{est}$ given in Eq. (2.24). A non-zero $\mathbf{C}_m^{-1}$ is analogous to a damped least squares solution. In practice, the statistical formulation allows us to formally constrain our solution based on our prior knowledge.

2.6.2 Case of weak non-linearity

In the case of weak non-linearity, the function $g(\mathbf{m})$ can be linearized around a reference model $\mathbf{m}_p$ as

$$g(\mathbf{m}) \cong g(\mathbf{m}_p) + \mathbf{G}_0(\mathbf{m} - \mathbf{m}_p), \tag{2.157}$$

where

$$G_0^{i\alpha} = \left(\frac{\partial^2 gi}{\partial \mathbf{m}^\alpha}\right)_{m=m_p}, \tag{2.158}$$

for the ith data point and αth model parameter. Thus an analysis similar to that described in the preceding section can be carried out to obtain a solution of the problem (Tarantola 1987).

2.6.3 Quasi-linear case

When the function $g(\mathbf{m})$ is still quasi-linear inside the significant region of the PPD, the linearization, as described earlier, is no longer acceptable, and the right strategy is to obtain the maximum of $\sigma(\mathbf{m}|\mathbf{d}_{obs})$, say $\mathbf{m}_M$, using some iterative algorithm and use a linearization around $\mathbf{m}_M$ for estimating the posterior covariance matrix. The point $\mathbf{m}_M$ that maximizes the PPD in Eq. (2.145) clearly minimizes the error functions given by Eq. (2.146). Several gradient methods can be employed to iteratively find a minimum of the error function (Tarantola 1987; Press *et al.* 1989). Some of these are Newton's method, the steepest-descent method, the preconditioned Newton method, the conjugate gradient (with preconditioning), the variable metric or quasi-Newton methods, and the simplex method.

2.6.4 Non-linear case

When the relationship between data and model is highly non-linear, the resulting error function may be characterized by several minima. The resulting PPD will be multimodal. Thus use of linear or iterative-linear methods will not be successful, and we will require using more computationally intensive global optimization algorithms.

2.7 Summary

In this chapter we reviewed direct, linear, and iterative-linear inverse methods in geophysics. Most direct methods are not robust for incomplete data in the presence of noise and are consequently not very useful. Therefore, model-based inversion methods have gained much popularity. Based on the nature of the forward problem, we followed the commonly accepted definition (e.g., Tarantola 1987) to classify inverse problems into linear, weakly non-linear, quasi-linear, and highly non-linear problems. These definitions may at times be somewhat misleading in that it is very difficult to say under what condition a problem can be classified as a weakly non-linear or quasi-linear problem. The facts are straightforward in that we know that most forward modeling operators are non-linear, and the success of an inverse method is usually judged by how well we explain the observed data given their uncertainty. Thus a probabilistic theory as developed by Tarantola or based on Bayes' rule seems to be adequate to describe the inverse problem. The solution to such problems, however, is the most difficult aspect. If a problem can be linearized – i.e., if, under some restrictive set of conditions, a forward modeling operator can be reduced to a linear operator – the solution becomes fairly straightforward. Otherwise, one needs to at least try to use gradient-based methods in solving the inverse problem. The gradient-based methods suffer from the following limitations:

• Gradient-based methods will often find a minimum in the close neighborhood of the starting solution. In situations where the error function may have several secondary minima, one needs to either start very close to the global minimum or try several starting points hoping that one of them will lead to the best solution.
• Most of these methods require that a gradient and sometimes a curvature (second-derivative) matrix be computed at each point during the iteration process. Although for some geophysical applications analytical expressions for the gradient and curvature can be derived, it is not possible in general. For example, for wave-propagation problems in laterally varying media, the forward calculation is done most accurately with methods such as finite differences. In such a case, the gradient can be computed numerically – a computationally formidable task. One alternative, which is becoming increasingly popular, is to employ an adjoint method to compute the gradient of the cost function.
• Some of these methods require computing the inverse of a large matrix. Fortunately, in many situations, several known methods such as Cholesky decomposition can be used for this purpose. Inverting a large matrix, even with well-known techniques, is not necessarily a very easy computational task.

Using a statistical framework, the solution of an inverse problem is described by the PPD. If the forward modeling is non-linear, clearly the PPD is not Gaussian, and the analysis of the solution is not straightforward. Even in the case of a highly non-linear problem, if one can start at a point very close to the peak of the PPD, the MAP point can be reached by gradient methods. But, using this approach, one would still only approximate the PPD by a Gaussian around the MAP point, even though the true PPD may be highly complicated in shape. The only way to truly estimate the PPD is to evaluate the error function at each point in model space, which is in general an impossible task. Therefore, practical ways need to be sought that can sample several points near the peak of the PPD such that the most significant part of the PPD can be approximately constructed by direct sampling rather than fitting a presumed shape around the MAP point.

3

Monte Carlo methods

In Chapter 2 we discussed linear, weakly non-linear, and quasi-nonlinear inverse problems and showed how they can be solved by several methods, such as the least squares, singular-value decomposition (SVD), etc. We noted that Tarantola's formulation of the inverse problem is very general and showed how it can be used for the special cases of linear and quasi-linear inversion problems under the assumption of Gaussian priors in data, model, and theory. One of the primary goals of inversion is to find the minimum of an error function. Due to the fact that such an error function may have several minima of different heights, a gradient-based non-linear approach may not be the most suitable method of inversion unless the reference model is adequately close to the true solution. Thus, in such a situation, we are left with the choice of a grid-search method or a Monte Carlo–type random-search technique. In fact, applications of such methods or combinations of these global and local methods (discussed in Chapter 2) to several geophysical problems have been reported in a few published articles. In this chapter we describe the grid-search and Monte Carlo methods with geophysical examples from published work in seismology. A review of the Monte Carlo method and its application to geophysical inversion can be found in Sambridge and Mosegaard (2002).

3.1 Enumerative or grid-search techniques

Enumerative or grid-search methods of inversion involve the systematic search through each point in a predefined model space to locate the best-fit models. For most geophysical problems, the model space is very large, and the forward calculation is slow. Therefore, use of grid-search methods for most geophysical applications is far from being practical. However, for simple problems with a few model parameters, this type of technique has been used successfully. For example, Walter (1993) used a grid-search technique in estimating the earthquake source parameters (i.e., strike, dip, and rake of the fault plane and the seismic moment) for the June 29, 1992, Little Skull mountain earthquake using complete regional seismic waveforms at a single station. Three-component displacement seismograms filtered between 50- and 15-second periods were used in the inversion. The seismic velocity model for the source-receiver path was assumed known from previous studies, and synthetic reflectivity seismograms were computed for six fundamental couples. Responses from these couples were combined to generate three-component seismograms due to any arbitrary mechanism (fault) at a specific azimuth (Aki and Richards 1980). This requires searching through different values of strike, dip, and rake so as to obtain synthetic seismograms that match the data the best. Note that the time-consuming reflectivity calculation needed to be done only once because it was assumed that the velocity function was known. The second step of the forward calculation, which requires combining the responses, is very rapid, and therefore, a grid-search procedure could be implemented. Walter (1993) used a two-step grid-search method for the source parameter estimation. The initial search was carried out over 10-degree increments of strike, dip, and rake. The solutions are shown in Figure 3.1. We observe two well-defined minima of nearly the same height; they correspond to the fault plane and the auxiliary plane. The two minima are near strike = 50 degrees; dip = 50 degrees; rake = −100 degrees; and strike = 220 degrees; dip = 40 degrees; rake = −80 degrees. Next, a grid search was applied in 1-degree increments near the first minimum to obtain: the moment $M_o = 4.1 \times 10^{17}$ N-m, strike = 35 degrees; dip = 50 degrees; and rake = −87 degrees. The synthetic seismograms for this mechanism provided a very good fit to the data (Figure 3.2).

Another excellent example of a grid-search technique to source estimation from broadband seismograms can be found in Zhao and Helmberger (1994). These authors introduced a novel technique that allows for better use of the entire broadband record when only imperfect Green's functions are available. The method used an average of L_1 and L_2 norms as a function of the four source parameters to define an error function whose minimum is found by the grid-search method. This example clearly demonstrated that prior information can be easily incorporated into the grid-search method.

In most of the grid-search inversions for earthquake source parameters, no attempt has been made to estimate the uncertainty in the results. This is a trivial task given that the error or the objective function is known at each point. Thus almost the entire posterior probability density function (PPD) can be computed, the marginal pdfs can be derived, and the posterior covariance matrix can easily be derived. One example of estimating these in earthquake location problems can be found in Tarantola and Valette (1982a).

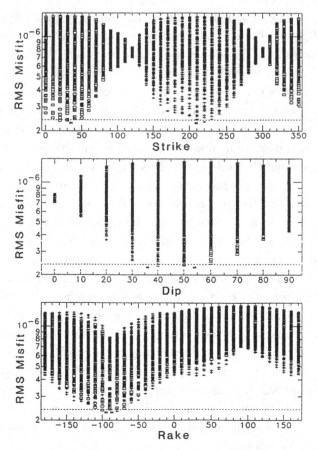

Figure 3.1 Example of grid-search inversion for earthquake source parameters. The RMS misfit values are shown for different values of strike, dip, and rake. Each point represents the error of one of the 12,960 mechanisms tested in the coarse (10-degree intervals) grid search. Strikes between 130 and 300 degrees have been given a different symbol to clearly show the two minima, one for each plane in the mechanism. (From Walter, 1993.)

3.2 Monte Carlo inversion

Monte Carlo methods are pure random-search methods in which models are drawn uniformly at random and tested against data. In a Monte Carlo inversion, each model parameter is allowed to vary within a predefined search interval (determined a priori). Thus, for each model parameter m_i, we define

$$m_i^{\min} \leq m_i \leq m_i^{\max}.$$

A random number is drawn from a uniform distribution $U[0, 1)$ and is then mapped into a model parameter. As an example, assume that the random number is rn; then the new model parameter value can be given as

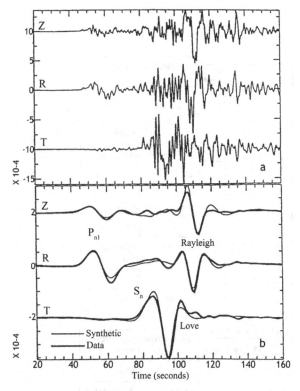

Figure 3.2 (*a*) Vertical (*Z*), radial (*R*), and transverse (*T*) component displacement seismograms recorded at KNB for the main shock. (*b*) The same traces bandpass-filtered between 50- and 15-second periods. Synthetic seismograms were calculated for the best-fit mechanism. (From Walter 1993.)

$$m_i^{\text{new}} = m_i^{\text{min}} + (rn)(m_i^{\text{max}} - m_i^{\text{min}}).$$ (3.1)

A new random model vector can be generated by random perturbation of a specific number of model parameters in the model vector. Synthetic data are then generated for the new model and compared with observations. The model is accepted deterministically based on an acceptance criterion that determines how well the synthetic data compare with the observations. The generation-acceptance/rejection process is repeated until a stopping criterion is satisfied. A commonly used stopping criterion is given by the total number of accepted models.

Press (1968) applied Monte Carlo inversion to determine earth models using as data: 97 eigenperiods of the earth's free oscillations, the travel times of compressional and shear waves, and, the mass and moment of inertia of the earth. The inversion derived the distribution of compressional-wave (α) and shear-wave (β) velocities and density (ρ) in the earth. Press's algorithm (taken from Press, 1968) is shown in Figure 3.3. As described earlier, the random selection

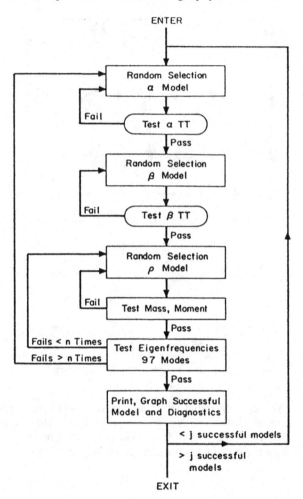

ENTER

Random Selection
α Model

Fail

Test α TT

Pass

Random Selection
β Model

Fail

Test β TT

Pass

Random Selection
ρ Model

Fail

Test Mass, Moment

Pass

Fails < n Times
Fails > n Times

Test Eigenfrequencies
97 Modes

Pass

Print, Graph Successful
Model and Diagnostics

< j successful models

> j successful
models

EXIT

Figure 3.3 Flow diagram of the Monte Carlo inversion procedure of Press. (From Press 1968.)

of each model parameter was prescribed to fall within upper and lower bounds that were different for different model parameters. The search ranges used by Press (1968) were broader than the bounds formed by most previously suggested models at that time. Although in computing the travel times, mass, moment of inertia, and eigenperiods, the earth model was discretized into 88 points, the model was allowed to vary only at 23 points (a total of 69 model parameters), and the remaining values were obtained by linear interpolation. Approximately five million models were tested, of which only six met all the constraints. It was found that the compressional-wave velocity did not affect the eigenperiod test, and therefore, the β and ρ distributions for the mantle are shown in Figure 3.4. Of the six models, three models were eliminated because of the implausibility of

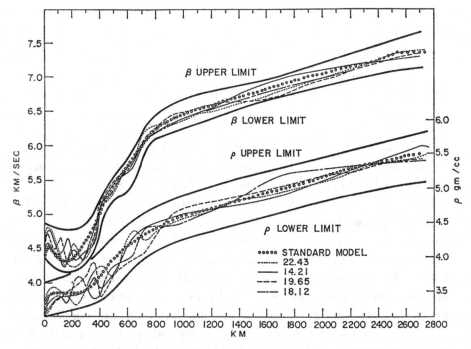

Figure 3.4 Results from the Monte Carlo inversion of Press (1965). Shear velocity (β) and density (ρ) distributions for the mantle for the standard model and for three successful Monte Carlo models. Model number also gives the increase in core radius in kilometers. Permissible β, ρ region shown by heavy curves. Model 14.21 failed the S_2 test but is of interest because of excellent fit of higher modes. (From Press 1968.).

zero density gradient in the deep mantle. The approach taken by Press (1968) is remarkable in that he was first to address the issue of uniqueness by direct sampling from the model space.

Several approaches can be taken to make Monte Carlo inversion more practical in many applications. For example, we may be able to find reasonable answers if the search space is reduced. Also, we may be able to describe the search window by means of a pdf (possibly a Gaussian with small variance) rather than a uniform distribution, based on prior information. If no prior information is available to narrow the search width, the best we can do is to introduce a new factor called the *search increment* to define model parameter resolution. Thus, for each model parameter m_i, we define a Δm_i such that the total number of possible values that the model parameter can take is given by

$$N_i = \frac{m_i^{\mathrm{max}} - m_i^{\mathrm{min}}}{\Delta m_i}. \tag{3.2}$$

Next, a random number rn drawn from $\mathbf{U}[0, 1]$ is mapped to an integer (say irn) between 1 and N; i.e.,

$$irn = N_i \cdot rn, \tag{3.3}$$

and then we compute the new model parameter as

$$m_i = m_i^{\min} + (irn) \cdot \Delta m_i. \tag{3.4}$$

For large values of Δm_i, the resolution will be poor, but it may be easier to reach a solution near the best model. Similarly, for small values of Δm_i, we may expect good resolution, but it may take enormous computing time.

One Monte Carlo approach that is probably more efficient than Press's (1968) approach can be found in the work by Wiggins (1969), who attempted to derive a suite of P-wave velocity models of the upper 1,000 km of the mantle that explain the observations of $dt/d\Delta$ and T versus Δ, where T is the travel time and Δ is the epicentral distance. Wiggins' (1969) approach was different from that of Press (1968) in that he not only picked a velocity bounded within certain limits for each boundary, but the velocity gradient (slope) and the change in slope also were bounded. Thus velocity models were drawn not from a purely uniform distribution but from a pdf such that all three criteria described earlier were satisfied. The objective was to find a representative sample of all possible velocity models that agree equally well with the travel-time observations. In order to increase the computational speed, Wiggins (1969) iterated the velocity models in a top-down fashion. That is, he first perturbed the shallowest part of the model that affected the shallow rays. Once he found a model that explained travel-time data from the shallowest part, the model in that part was fixed, and the deeper part of the model was perturbed, and so on. Proceeding in this manner, a set of fifty models was obtained that passed all the observational tests. These models were then used to obtain bounds on the compressional-wave velocity and thickness of different layers.

There are, however, two issues related to the Monte Carlo method that need to be addressed. They are

- How can the set of successful models be used to estimate the uncertainty or resolution of the derived result?
- How does one know if one has indeed obtained a representative set of models that agree with the observations? In other words, what is the criterion to stop further sampling?

These two questions are related, and Kennett and Nolet (1978) were the first to address them. They showed how the concepts of resolution originally developed for linear problems and extended to locally linear regions can be extended to the non-linear region as well. We will closely follow Kennett and Nolet (1978) to show this. Let $\mathbf{m}^1$, $\mathbf{m}^2$, $\mathbf{m}^3$, ..., $\mathbf{m}^L$ be a set of successful models each consisting of NM model parameters m_i, $i = 1, 2, 3, ..., NM$. The average or the *mean* model $<\mathbf{m}>$ will be given by

$$\langle \mathbf{m} \rangle = \frac{1}{L} \sum_{i=1}^{L} \mathbf{m}^i. \tag{3.5}$$

Thus the deviations of successful models from the mean model can be obtained from the following relation:

$$\Delta \mathbf{m}^i = \mathbf{m}^i - \langle \mathbf{m} \rangle. \tag{3.6}$$

We can also construct a new matrix $\mathbf{P}$ as

$$\mathbf{P} = \frac{1}{L} \sum_{i=1}^{L} \left(\mathbf{m}^i - \langle \mathbf{m} \rangle \right) \left(\mathbf{m}^i - \langle \mathbf{m} \rangle \right)^T. \tag{3.7}$$

Kennett and Nolet (1978) showed that if $\mathbf{V}_k$ are the eigenvectors of the asymptotic value of $\mathbf{P}$ (i.e., in the limit, $L \to \infty$), the resolution matrix $\mathbf{R}$ will be given as

$$\mathbf{R} = \sum_{k=1}^{K} \mathbf{V}_k \mathbf{V}_k^T, \tag{3.8}$$

where K is the number of eigenvalues of $\mathbf{P}$. However, in practice, L is finite, and $\mathbf{V}_p$ needs to be computed from the asymptotic value of $\mathbf{P}$. This can be achieved numerically by evaluating Eq. (3.8) each time five or ten successful models have been found, and the inversion can be stopped when there is no significant variation in the estimated value of the resolution matrix (Kennett and Nolet 1978). These ideas are very simple but are very important and useful.

3.3 Hybrid Monte Carlo–linear inversion

Monte Carlo methods can work for highly non-linear inverse problems. However, as demonstrated earlier, they are computationally very expensive. Therefore, several different schemes have to be adopted for their implementation. One approach is to use prior information to define a prior pdf from which to sample the models such that the search space is narrow. Another approach is to use a large search increment and accept a low-resolution result. The latter approach can also be implemented by using data from which only results with large uncertainty can be expected; i.e., for such a data set, the global minimum region of the error or misfit surface is broad. For example, use of travel times alone in seismic inversion results in velocity models that are not as well constrained as those using the whole seismograms, i.e., using information on seismic amplitudes, travel times, and waveforms. The forward problem of travel-time calculation is also fast compared with synthetic seismogram generation, and thus travel-time inversion problems can be solved by the Monte Carlo method, whereas the full-waveform inversion problem can be solved by one of the gradient methods. This is the approach taken by Cary and Chapman (1988), who used a two-step procedure in seismic waveform inversion. In the first step of their algorithm, they used a Monte Carlo search through a large model space defined from poor prior knowledge of the velocities. The data for this step of the inversion were travel-time observations from the seismograms and were used to eliminate poorly fitting models. This resulted in a model that was within the valley of the global minimum. In the next step, the immediate neighborhood of the global minimum was explored with constrained least squares inversion. At this step, the complete seismograms (i.e., travel times, waveforms, and amplitudes) were used in the inversion using WKBJ seismograms. In addition to deriving the best-fit model, Cary and Chapman (1988) also constructed marginal pdfs for the model parameters and their covariances. Their analysis showed the correlations among different model parameters.

3.4 Directed Monte Carlo methods

As described earlier, Monte Carlo methods evaluate the error function at many points that are randomly selected in model space. Unfortunately, pure Monte Carlo methods are too expensive to employ for most interesting problems. Thus the fundamental question is: Can we develop more efficient methods to sample model space? In particular, we want to retain the following features: The sampling must involve a *large* degree of randomness so as not to prebias our results. For example, we do not want to get trapped in a local minimum of the error function. Further, since we cannot afford a pure Monte Carlo search (or an enumeration of all possible models), we would like to direct our search of the model space toward sampling the good models.

Directed Monte Carlo methods, such as simulated annealing and genetic algorithms that are the main subject of this book, specifically address these issues. Both these methods use random walks, but unlike a pure Monte Carlo method, they use a transition probability rule to guide their search and have been proven to be superior to pure Monte Carlo methods in many applications. We describe these techniques in detail in the following chapters.

4

Simulated annealing methods

One of the major goals of geophysical inversion, whether described in terms of probability theory or by more classical approaches, is searching for the minimum of an error function $E(\mathbf{m})$, where $\mathbf{m}$ is a model vector. Local optimization methods such as iterative linear methods compute an update to the current model by assuming that the error surface has a well-defined minimum. Any update to the current model is only accepted if the error computed for the updated model is less than the error for the previous model. Clearly, these approaches will fail when the error surface has several peaks and troughs. These methods will always find the minimum closest to the starting model, and that is why they are often called *greedy algorithms*.

Simulated annealing (SA) is an alternative method for finding the global minimum of a function $E(\mathbf{m})$. The method has been found to be useful in a variety of optimization problems (e.g., Kirkpatrick *et al.* 1983; Geman and Geman 1984; van Laarhoven and Aarts 1987; Aarts and Korst 1989; etc.) that involve finding optimal (minimum or maximum) values of a function of a very large number of independent variables. In many of these applications, the function to be minimized is called the *cost function* (which is similar to the error function defined in Chapter 2) by analogy with the traveling salesman problem, in which the optimal route (with minimum cost) is searched for a salesman who can travel through each city once and finally return to the starting point.

Geophysical inverse problems also involve finding the minimum of an error function $E(\mathbf{m})$ that is usually a function of a large number of variables, i.e., model parameters. Consequently, SA has been used successfully in many geophysical inverse problems (e.g., Rothman 1985, 1986; Basu and Frazer 1990; Sen and Stoffa 1991; etc.).

The basic concepts of SA are borrowed from problems in statistical mechanics that involve the analysis of the properties of a large number of atoms in samples of liquids or solids. It draws an analogy between the parameters (model parameters) of an optimization problem and particles in an idealized physical system. A physical annealing process occurs when a solid in a heat bath is initially heated by increasing the temperature such that all the particles are distributed randomly in a liquid phase. This is followed by slow cooling such that all the particles arrange themselves in the low-energy ground state where crystallization occurs. The optimization process involves simulating the evolution of the physical system as it cools and anneals into a state of minimum energy. In the terminology of stochastic processes, each configuration of particles is referred to as a *state*. At each temperature, the solid is allowed to reach thermal equilibrium, where the probability of being in a state i with energy E_i is given by the following Gibbs or Boltzmann pdf:

$$P(E_i) = \frac{\exp\left(-\dfrac{E_i}{KT}\right)}{\displaystyle\sum_{j \in S} \exp\left(-\dfrac{E_j}{KT}\right)} = \frac{1}{Z(T)} \exp\left(-\frac{E_i}{KT}\right), \tag{4.1}$$

where the set S consists of all possible configurations, K is Boltzmann's constant, T is the temperature, and $Z(T)$, the partition function, is given by

$$Z(T) = \sum_{j \in S} \exp\left(\frac{E_j}{KT}\right) \tag{4.2}$$

The temperature is reduced gradually after thermal equilibrium has been reached such that in the limit $T \rightarrow 0$, the minimum energy state becomes overwhelmingly probable. The key to this process is the requirement of equilibrium. If the cooling is done too rapidly (quenching), the material will freeze at a local minimum of E, forming a glass. However, if the melt is cooled very slowly (annealing), then it will eventually freeze into an energy state that is at or very close to the global minimum of E, forming a crystal.

In optimization problems such as those that occur in geophysical inverse problems, the energy function is identified with the error function $E(\mathbf{m})$. The error is a function of the possible geophysical models and takes on different values for different configurations or models. We are interested in finding the state (or model) $\mathbf{m}$ for which this error is minimum. SA is implemented using a computer algorithm that mimics the physical annealing process to find the global minimum energy state. Before discussing how this is actually done or what kind of problems one encounters, we illustrate graphically why SA can often be expected to be more useful than the greedy algorithms.

First, for geophysical applications, we set the Boltzmann constant K equal to 1 and rewrite the Gibbs pdf in Eq. (4.1) as follows:

$$P(\mathbf{m}_i) = \frac{\exp\left[-\dfrac{E(\mathbf{m}_i)}{T}\right]}{\displaystyle\sum_{j \in S} \exp\left[-\dfrac{E(\mathbf{m}_j)}{T}\right]}, \tag{4.3}$$

where T is a control parameter that has the same dimension as that of energy or error. For discrete inverse problems, if we assume that there exist N model parameters such that each model parameter can take M possible values, the set S consists of M^N possible model configurations. In typical geophysical inverse problems, we can easily have as many as 50^{50} models.

For illustration purposes, let us assume that we have only two variables x and y and consider a function of the two variables of the type

$$f(x, y) = \text{sgn}\left(\frac{\sin x}{x}\right)\left(\left|\frac{\sin x}{x}\right|\right)^{\frac{1}{4}} \text{sgn}\left(\frac{\sin y}{y}\right)\left(\left|\frac{\sin y}{y}\right|\right)^{\frac{1}{4}}. \tag{4.4}$$

This function has its global maximum value of 1.0 at $x = 0$, $y = 0$. However, it also has several secondary maxima. To cast this into a problem of minimization, we define the error function

$$E(x, y) = \left[1 - f(x, y)\right]^2, \tag{4.5}$$

which has a global minimum value of 0.0 at $(0, 0)$ and several secondary minima (Figure 4.1a). Now, substituting the values of $E(x, y)$ in Eq. (4.3) for different values of T (e.g., $T = 10.0$, $T = 1.0$, $T = 0.1$, $T = 0.01$), we can construct the Gibbs pdf at different temperatures, as shown in Figure 4.1b–e. The effect of this temperature T is to exaggerate or accentuate the differences between different values of the error function. Figure 4.1b–e clearly shows that at high temperatures, the pdf is nearly uniform; i.e., every model (x, y) has nearly equal probability. As the temperature is gradually lowered, the peaks become distinguishable, and at very low temperatures (e.g., 0.01), the peak of the pdf corresponding to the global minimum of the function clearly dominates.

This example helps us to understand the role of temperature and suggests how SA might be used to locate the global minimum of a multidimensional function. For a real multidimensional problem, however, this is not a trivial task. Examination of the Gibbs distribution in Eq. (4.3) shows that to construct the pdf, we have to evaluate the partition function in the denominator of Eq. (4.3). This requires that the error function be evaluated at each point. On the other hand, if $E(\mathbf{m})$ is known at each point in the model space, there is no need to use SA.

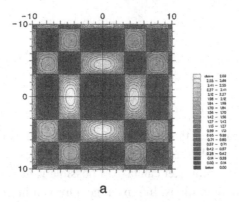

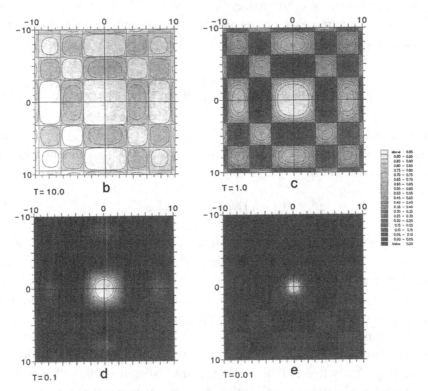

Figure 4.1 (*a*) Plot of the error function defined in Eqs. (4.4) and (4.5). Note the existence of the global minimum at (0, 0) and several secondary minima. This error function was used in computing the Gibbs pdf (Eq. 4.3) at different temperatures. (*b*) $T = 10.0$. (*c*) $T = 1.0$. (*d*) $T = 0.1$. (*e*) $T = 0.01$. At high temperatures, the pdf is nearly uniform. At very low temperatures, the peak of the pdf corresponding to the global minimum of the error function dominates.

Table 4.1. *List of simulated annealing algorithms*

Metropolis algorithm
Heat bath algorithm
SA without rejected moves
Fast simulated annealing
Very fast simulated reannealing
Mean field annealing

Several computer algorithms have been proposed to avoid computing error functions at each point in model space and yet achieve an approximation to the Gibbs
pdf asymptotically. Several of these are listed in Table 4.1. A good review of SA
with a list of references can be found in Ingber (1993).

4.1 Metropolis algorithm

In the early days of digital scientific computing, Metropolis *et al.* (1953) introduced an algorithm for simulating the evolution of a solid in a heat bath to thermal
equilibrium. Since the work of Kirkpatrick *et al.* (1983), the Metropolis algorithm
with annealing has been used in a wide variety of applications. The algorithm as
used currently is described schematically in Figure 4.2. Given a starting model $\mathbf{m}_i$
with energy $E(\mathbf{m}_i)$, a small perturbation to $\mathbf{m}_i$ is made to obtain a new model $\mathbf{m}_j$
given by

$$\mathbf{m}_j = \mathbf{m}_i + \Delta\mathbf{m}_i, \tag{4.6}$$

with energy $E(\mathbf{m}_j)$. Metropolis *et al.* (1953) generated the new state (model) by the
displacement of a particle (model parameter). However, any general perturbation
of the form given in Eq. (4.6) is applicable. If ΔE_{ij} is the difference in the energy
between the two states, i.e.,

$$\Delta E_{ij} = E\left(\mathbf{m}_j\right) - E\left(\mathbf{m}_i\right), \tag{4.7}$$

then whether or not the new model is accepted is decided based on the value of
ΔE_{ij}. If $\Delta E_{ij} = 0$, the new model is always accepted as in the local optimization
algorithms described in Chapter 2. However, if $\Delta E_{ij} > 0$, then the new model is
accepted with the probability

$$P = \exp\left(-\frac{\Delta E_{ij}}{T}\right), \tag{4.8}$$

where T is the temperature. The preceding acceptance rule (Eq. 4.8) is known as
the *Metropolis criterion*. If the generation-acceptance process using the preceding

Metropolis Algorithm

start at a random location with $\mathbf{m}_0$ with energy $E(\mathbf{m}_0)$

loop over temperature (T)

- loop over number of random moves/temperature
- • calculate $E(\mathbf{m}_1)$ for a new model $\mathbf{m}_1$
- • $\Delta E = E(\mathbf{m}_1) - E(\mathbf{m}_0)$
- • $P = \exp\left(-\dfrac{\Delta E}{T}\right)$

    if $\Delta E \leq 0$, then
- • • $\mathbf{m}_0 = \mathbf{m}_1$
- • • $E(\mathbf{m}_0) = E(\mathbf{m}_1)$
- • endif
- • if $\Delta E > 0$, then
- • • draw a random number $r = U[0, 1]$
- • • if $P > r$, then
- • • • $\mathbf{m}_0 = \mathbf{m}_1$
- • • • $E(\mathbf{m}_0) = E(\mathbf{m}_1)$
- • • end if
- • end if
- end loop

end loop

Figure 4.2 A pseudo-Fortran code for a Metropolis simulated annealing algorithm.

criteria is repeated a large number of times at each temperature, it can be shown that thermal equilibrium can be attained at each temperature. If the temperature is lowered slowly following a cooling schedule such that thermal equilibrium has been reached at each temperature, then in the limit as the temperature approaches zero, the global minimum energy state can be reached. Theoretical cooling schedules that statistically guarantee convergence to the global minimum can be derived (e.g., Geman and Geman 1984).

In a local search method, we start with a reference model, and a new model is accepted if and only if $\Delta E_{ijl} = 0$; i.e., it always searches in the downhill direction. However, in SA, as described by the Metropolis algorithm, every model has a finite probability of acceptance even though $\Delta E_{ij} > 0$. Thus local methods can get trapped in a local minimum that may be in the close neighborhood of the starting model, whereas SA (not being so *greedy*) has a finite probability of jumping out of local minima. As the temperature approaches zero, however, only the moves that show improvement over the preceding trial are likely to be accepted, and in the limit $T \to 0$, the algorithm reduces to a greedy algorithm.

The computational algorithm is described in Figure 4.2. In Figure 4.3 we show one of the many ways of generating a random perturbation. Starting with a model $\mathbf{m}_0$ with eight parameters such that each parameter can have eight possible values (m_{ij}, I = model parameter index, j = one of the possible values for each model parameter), a random move can be made by perturbing the first model parameter from m_{14} to m_{17}. The error for the new model is computed, and the Metropolis criterion is evaluated to decide whether or not this move should be accepted. After a certain number of moves at a temperature, the temperature is lowered according to a *cooling schedule*, and the process is repeated until convergence. In this case, convergence is achieved when the energy remains unchanged for several iterations. In practical applications, defining the initial temperature is problem-dependent.

4.1.1 Mathematical model and asymptotic convergence

The process of generation and acceptance of models according to the Metropolis rule can be modeled with finite Markov chains, as described in Chapter 1. Recall that a Markov chain is a sequence of trials in which the probability of the outcome of a given trial depends only on the outcome of the preceding trial and not on any other. Clearly, in Metropolis SA, the probability of the outcome of a trial depends only on the outcome of the preceding trial through the Metropolis criterion. We also showed earlier that for Markov chains that satisfy the properties of irreducibility and aperiodicity, there exists a unique stationary distribution such that the probability of the occurrence of a model is independent of the starting model. To determine the form of the stationary distribution and whether such a stationary distribution indeed exists for a Metropolis SA modeled with Markov chains, we need to examine the transition probability matrix and the irreducibility and aperiodicity properties of the chain. The following section has been adopted largely from Aarts and Korst (1989) and van Laarhoven and Aarts (1987).

For Metropolis SA, we can write the transition probability P_{ij} for transition from model i to model j as a product of two probabilities, namely, G_{ij} (the generation probability) and A_{ij} (the acceptance probability) such that

$$P_{ij}(k) = P_{ij}(T_k) = \begin{cases} G_{ij}(T_k) A_{ij}(T_k) & \text{if } i \neq j, \\ 1 - \sum_{l \neq i} P_{il}(T_k) A_{ij} & \text{if } i = j, \end{cases} \tag{4.9}$$

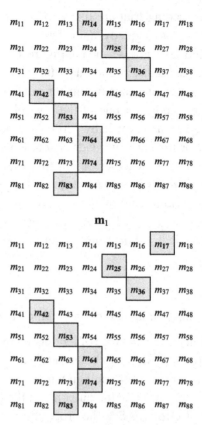

Figure 4.3 Metropolis SA starts with a model $\mathbf{m}_0$. The top figure shows a starting model configuration for a model with eight model parameters such that each one is allowed to have eight possible values. The shaded boxes are the values of the model parameters comprising a model. Next, a new model is obtained by perturbing $\mathbf{m}_0$. Here this is done (bottom) by only changing the first model parameter value from m_{14} to m_{17}. The new model is accepted or rejected based on Metropolis criterion (Eq. 4.8).

where P_{ij}, G_{ij}, and A_{ij} are the elements of matrices $\mathbf{P}$, $\mathbf{G}$, and $\mathbf{A}$ called the *transition*, *generation*, and *acceptance probability matrices*, respectively. The time step is k, and we represent all elements as a function of the temperature T_k, which depends generally on the time step or iteration number k. Clearly, the $\mathbf{P}$ matrix thus defined is a stochastic matrix because $P_{ij} = 1$ [see Chapter 1, Eq. (1.64)].

In classical Metropolis SA, the generation probability G_{ij} is uniform; i.e., models are drawn from a uniform distribution at random. The acceptance probability A_{ij} is, of course, temperature-dependent and is given by the Metropolis criterion. This can be written as (Aarts and Korst 1989)

$$A_{ij}(T_k) = \exp\left(-\frac{\left[E(\mathbf{m}_j) - E(\mathbf{m}_i)\right]^+}{T_k}\right), \tag{4.10}$$

where $a^+ = a$ if $a > 0$ and $a^+ = 0$ otherwise, where $a = E(\mathbf{m}_j) - E(\mathbf{m}_i)$. Thus a finite (but temperature-dependent) probability of acceptance occurs for worse models $E(\mathbf{m}_j) > E(\mathbf{m}_i)$, and better models $E(\mathbf{m}_j) < E(\mathbf{m}_i)$ are always accepted. Now that we have defined the transition matrix $\mathbf{P}$, we need to examine whether the Markov chain is irreducible and aperiodic.

4.1.1.1 Irreducibility

Irreducibility requires that for each pair of models i, j there is a positive probability of reaching j in a finite number of trials n; i.e.,

$$P_{ij}^n > 0. \tag{4.11}$$

The n-step transition matrix can be derived by using the Chapman–Kolmogorov equation (Eq. 1.74). Since P_{ij} is the probability of transition from state i to state j in one time step, we have from the Chapman–Kolmogorov equation

$$\begin{aligned}
P_{ik_2}^2 &= \sum_{k_1} P_{ik_1} P_{k_1 k_2}, \\
P_{ik_3}^3 &= \sum_{k_2} P_{ik_2}^2 P_{k_2 k_3} = \sum_{k_1} \sum_{k_2} P_{ik_1} P_{k_1 k_2} P_{k_2 k_3}, \\
P_{ik_4}^4 &= \sum_{k_1} \sum_{k_2} \sum_{k_3} P_{ik_1}^2 P_{k_1 k_2} P_{k_2 k_3} P_{k_3 k_4}, \\
P_{ij}^n(T) &= \sum_{k_1} \sum_{k_2} \cdots \sum_{k_{n-1}} P_{ik_1}(T) P_{k_1 k_2}(T) \cdots P_{k_{n-1} j}(T).
\end{aligned} \tag{4.12}$$

Now recall that each term P_{ik} is given by the product $G_{ik} A_{ik}$ (Eq. 4.9). We also know that $A_{ij} > 0$, for all i, j, and it is also justified to assume that each $G_{ij} > 0$ for the generation matrix. Therefore, each element P_{ik} in the sum is a positive number. Thus the sum should be greater than or equal to the individual elements in the sum. This means that

$$P_{ij}^n(T) \geq G_{il_1} A_{il_1}(T) G_{l_1 l_2} A_{l_1 l_2}(T) \cdots G_{l_{n-1} j} A_{l_{n-1} j}(T) > 0. \tag{4.13}$$

This proves irreducibility of the chain.

4.1.1.2 Aperiodicity

A Markov chain with period 1 is said to be *aperiodic*. Therefore, for every state i, we should have $P_{ii} > 0$. For two states i and j with different energies $E(\mathbf{m}_i)$ and $E(\mathbf{m}_j)$, we have, from Eq. (4.9),

$$P_{ii}(T) = 1 - \sum_{l \neq i} G_{il} A_{il}(T),$$

$$= 1 - G_{ij} A_{ij}(T) - \sum_{l \neq i,j} G_{il} A_{il}(T),$$

$$> 1 - G_{ij} - \sum_{l \neq i,j} G_{il}, \qquad \text{(Recall } A_{ij} < 1 \text{ and positive)} \tag{4.14}$$

$$> 1 - \sum_{l \neq i} G_{il} > 0.$$

This proves aperiodicity of the Markov chain defined by the transition matrix **P** as given by Eq. (4.9).

4.1.1.3 Limiting probability

In Section 1.13 we showed that for a finite, irreducible, and aperiodic homogeneous Markov chain, a unique stationary distribution exists that satisfies Eqs. (1.77) and (1.78). We just showed, following Aarts and Korst (1989), that Metropolis SA modeled as a time-homogeneous Markov chain is irreducible and aperiodic. It remains to identify a stationary distribution **q** that satisfies Eq. (1.78). Let us assume that the stationary distribution **q** is given by the Gibbs pdf such that each element of vector **q** is given by the following equation:

$$q_j = \frac{1}{Z(T)} \exp\left[-\frac{E(\mathbf{m}_j)}{T}\right], \tag{4.15}$$

where

$$Z(T) = \sum_{i \in S} \exp\left[-\frac{E(\mathbf{m}_i)}{T}\right], \tag{4.16}$$

and the stochastic matrix **P** is defined in Eq. (4.9). We now have (following Aarts and Korst 1989)

$$q_i P_{ij} = q_i G_{ij} A_{ij}$$

$$= \frac{G_{ij}}{Z(T)} \exp\left[-\frac{E(\mathbf{m}_i)}{T}\right] \exp\left[-\frac{E(\mathbf{m}_j) - E(\mathbf{m}_i)^+}{T}\right]$$

$$= \frac{G_{ij}}{Z(T)} \exp\left[-\frac{E(\mathbf{m}_j)}{T}\right] \exp\left(-\frac{\left[E(\mathbf{m}_i) - E(\mathbf{m}_j)\right] + \left[E(\mathbf{m}_j) - E(\mathbf{m}_i)\right]^+}{T}\right)$$

$$= \frac{G_{ij}}{Z(T)} \exp\left[-\frac{E(\mathbf{m}_j)}{T}\right] \exp\left(-\frac{\left[E(\mathbf{m}_i) - E(\mathbf{m}_j)\right]^+}{T}\right),$$

assuming $E(\mathbf{m}_i) > E(\mathbf{m}_j)$, or

$$q_i P_{ij} = G_{ij} \frac{1}{Z(T)} \exp\left[-\frac{E(\mathbf{m}_j)}{T}\right] A_{ji} = q_j G_{ij} A_{ji}.$$

Since G_{ij} is a uniform distribution, we can assume that $G_{ij} = G_{ji}$. Therefore, we have

$$q_i P_{ij} = q_j P_{ji},$$

which is the same as Eq. (1.78). This shows that the Gibbs distribution is indeed the stationary distribution for the Metropolis SA. This means that if the generation-acceptance process according to the Metropolis rule is repeated a large number of times, the models are distributed according to the Gibbs pdf. We have proved this for time-homogeneous Markov chains with generation and acceptance probabilities as defined by Eq. (4.9). A similar analysis for more general classes of acceptance and generation probabilities and for time inhomogeneous Markov chains can also be carried out (Anily and Federgruen 1987a, 1987b; Mitra *et al.* 1986).

4.2 Heat bath algorithm

The Metropolis algorithm, as described in the preceding section, can be considered a two-step procedure in which a random move is first made, and then it is decided whether or not the move should be accepted. Thus many of these moves will be rejected. Especially at low temperatures, the rejection-to-acceptance ratio is very high. Several algorithms have been proposed to remedy this situation – one of which is the *heat bath algorithm* (Rebbi 1984; Creutz 1984; Geman and Geman 1984; Rothman 1986).

Unlike the Metropolis algorithm, the heat bath algorithm is a one-step procedure that attempts to avoid a high rejection-to-acceptance ratio by computing the relative probabilities of acceptance of each trial move before any random guesses are made. The method simply produces weighted selections that are always accepted. The preceding statement may be somewhat misleading because a weighted guess is made only at substantial expense, as will become clear by analyzing the algorithm in detail. A schematic diagram of the heat bath algorithm is given in Figure 4.4.

Consider a model vector $\mathbf{m}$ consisting of N model parameters. Next, assume that each m_i can take M possible values. This can be obtained by assigning some lower (m_i^{min}) and upper (m_i^{max}) bounds, and conducting a search increment Δm_i for each model parameter such that

$$M = \frac{m_i^{max} - m_i^{min}}{\Delta m_i}. \tag{4.17}$$

Heat Bath Algorithm

start at a random location with $\mathbf{m}_0$ with energy $E(\mathbf{m}_0)$

loop over temperature (T)

- loop over number of iterations/temperature
- - loop over number of model parameters ($i=1, \dots, N$)
- - - loop over model values $j, j=1, \dots, M$
- - - - evaluate $E\,(\mathbf{m} \mid m_i = m_{ij})$ and calculate

$$\hat{P}_{ij} \propto \exp\left(-\frac{E\left(\mathbf{m} \mid m_i = m_{ij}\right)}{T}\right)$$

- - - end loop
- - - draw a j from the above distribution
- - end loop
- end loop

end loop

Figure 4.4 A pseudo-Fortran code for a heat bath simulated annealing algorithm.

Note that M can be different for different model parameters. However, without loss of generality, we will assume that M is the same for all model parameters. This results in a model space consisting of M^N models. At this stage it is convenient to represent model parameters by the symbol m_{ij}, $i = 1, 2, \dots, N$, $j = 1, 2, \dots, M$. The index i is the model parameter number, and j represents the discrete values of the model parameter. The algorithm is described graphically in Figure 4.5 with an example model consisting of eight model parameters each of which has eight possible values.

The algorithm starts with a random model, e.g., $\mathbf{m}_0$. Then each of the model parameters is visited sequentially. For each model parameter (say, i), the following marginal pdf is evaluated.

$$\hat{P}\left(\mathbf{m} \mid m_i = m_{ij}\right) = \frac{\exp\left[-\dfrac{E\left(\mathbf{m} \mid m_i = m_{ij}\right)}{T}\right]}{\displaystyle\sum_{k=1}^{M}\left[-\dfrac{E\left(\mathbf{m} \mid m_i = m_{ik}\right)}{T}\right]}, \tag{4.18}$$

where the sum is over M values that the model parameter is allowed to have and $E(\mathbf{m}|m_i = m_{ij})$ is the energy for a model vector $\mathbf{m}$ whose ith model parameter has a value m_{ij}. Then a value is drawn from the preceding distribution. Essentially, a

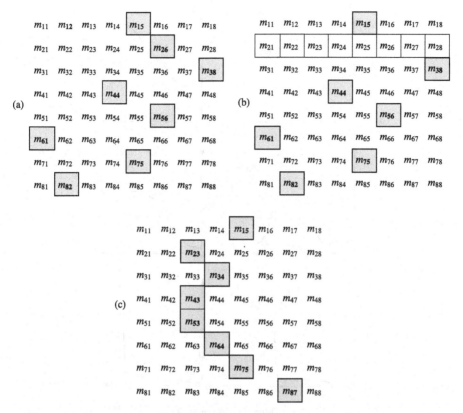

Figure 4.5 For a model with eight model parameters (each having eight possible values), the heat bath algorithm starts with a randomly chosen model shown by shaded boxes in (a). Each model parameter is then scanned in turn, keeping all others fixed. (b) Scanning through the model parameter m_2. Thus m_{26} is replaced with m_{23}, and we have a new model described by shaded boxes in (c). This process is repeated for each model parameter.

random number drawn from the uniform distribution $U[0,1]$ is mapped onto the pdf defined by Eq. (4.18). To do this, we first compute a cumulative probability C_{ij} from $\hat{P}\left(\mathbf{m}|m_i = m_{ij}\right) P(\mathbf{m} \mid m_i = m_i)$ as

$$C_{ij} = \sum_{k=1}^{j} \hat{P}\left(\mathbf{m}|m_i = m_{ik}\right), \quad j = 1, 2, \ldots, M. \tag{4.19}$$

Then we draw a random number r from the uniform distribution. At the point $j = k$, where $C_{ij} = r$, we select $m_{ij} = m_{ik} = m_i$ (in practice, the least integer k for which $C_{ij} > r$ is chosen). The procedure of drawing a model parameter from the

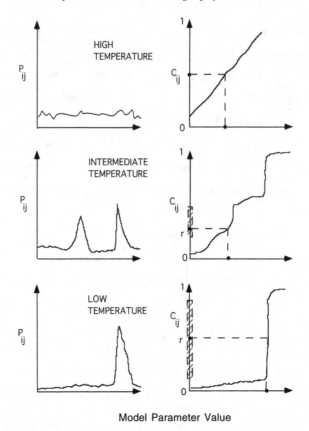

Figure 4.6 Model-generation probabilities (*left*) and their corresponding cumulative probabilities (*right*) at different temperatures. At high temperature, the distribution is nearly flat, and every model parameter value is equally likely. At intermediate temperature, peaks start to appear, and at low temperature, the probability of occurrence of the best solution becomes very high. A random number is drawn from a uniform distribution and is mapped to the cumulative distribution (*right*) to pick a model.

preceding distribution is illustrated graphically in Figure 4.6. At high temperatures, the distribution is nearly uniform, and therefore, every model parameter is nearly equally likely to be picked. At very low temperatures, only the peak corresponding to the model parameter that influences the error function the most dominates.

The new value of model parameter thus picked replaces the old value of the corresponding model parameter in the model vector, and the process is repeated sequentially (in any order) for each model parameter. After each model parameter is examined once, we may have a model that is different from the starting model. This constitutes one iteration of the algorithm that consists of evaluating the error function (or the forward calculation) $N \times M$ times. Unlike Metropolis SA, the heat bath does not require testing whether or not a model should be accepted after it has been generated, but this algorithm requires considerable computational expense. It has been reported that for problems with a very large number of variables, this may, however, be faster than Metropolis SA (Rothman 1985).

Scanning through each model parameter is repeated a large number of times at a constant temperature. Then the temperature is lowered according to a cooling schedule, and the process is repeated until convergence is achieved.

4.2.1 Mathematical model and asymptotic convergence

Although Metropolis SA and the heat bath algorithms appear to be entirely different search algorithms, it has been shown that essentially they do the same thing (e.g., Rothman 1986; Geman and Geman 1984). That is, the heat bath algorithm also attains a unique equilibrium distribution given by the Gibbs distribution after a large number of transitions at a constant temperature. This can be shown again by modeling the process by Markov chains (Rothman 1986). The first step in this is to identify the transition probability matrix and show that the Markov chain is irreducible and aperiodic. The following section on the asymptotic proof of convergence has been adopted largely from Rothman (1986).

4.2.1.1 Transition probability matrix

As discussed in the preceding section, each element of the transition probability matrix P_{ij} is the probability of transition from model i to model j. For a model vector **m** consisting of N model parameters such that each model parameter can take M possible values, the transition matrix **P** is an $M^N \times M^N$ matrix.

In a heat bath algorithm, each model parameter is visited sequentially. Therefore, at any time [before a model parameter value is drawn from the distribution in Eq. (4.17)], only M new models are accessible. Therefore, for one step through each model parameter, we can define a transition matrix, say, **P**(n), where n is the model parameter number. Thus there exist N different transition probability matrices **P**(1), **P**(2), ..., **P**(N). Note that one sweep through all the model parameters constitutes one time step of the Markov chain. Therefore, the transition probability from one complete iteration to another is given by the product of N matrices

$$\mathbf{P} = \mathbf{P}(1)\,\mathbf{P}(2)\,\mathbf{P}(3)\cdots\mathbf{P}(N). \tag{4.20}$$

Unlike the discussions of transition probabilities in Chapter 1 or in the Metropolis algorithm, we have two symbols for transition probabilities: One is **P**(n) with entries $P_{ij}(n)$ that corresponds to the transition matrix corresponding to a model parameter n, and the other is simply **P** with entries P_{kl} that correspond to transition from model vector k to model vector l through one step of the Markov chain. They are related by Eq. (4.20). This distinction is very important because it makes it possible to carry out the analysis shown in the following sections.

4.2.1.2 Irreducibility

The irreducibility property of a Markov chain implies that every model can be reached from every other model in a finite number of transitions with some positive probability. This may not appear obvious in the case of the heat bath algorithm, where each model parameter is visited sequentially. The proof can, however, be shown by close examination of the transition matrices **P**(n) for each parameter (Rothman 1986).

If in a set of models (or states) the models can communicate or one model can be reached from another, then the models are said to belong to the *same ergodic class*.

First, let us consider the example of a model with two parameters such that each one can take four trial values, i.e., for an m_{ij}, $i = 1, 2$ and $j = 1, 2, 3, 4$. Figure 4.7 shows that in this case there are sixteen possible configurations or models shown by dots. Now let us assume that we start with a random value for m_{2j} (say, m_{21}) and search through all possible values of m_{1j}; i.e., we search through four models $(m_{11}, m_{21})^T$, $(m_{12}, m_{21})^T$, $(m_{13}, m_{21})^T$, and $(m_{14}, m_{21})^T$ that belong to the same ergodic class. Thus, for four different starting values of m_{2j}, we will have four sets of ergodic classes. A similar analysis can also be carried out for m_2 with m_1 fixed. The ergodic classes of **P**(1) and **P**(2) are marked in Figure 4.7. Thus each transition matrix partitions the set of M^N (sixteen in the example in Figure 4.7) models into M^{N-1} (four for the example in Figure 4.7) ergodic classes. Notice also in the figure that every element in any ergodic class of **P**(1) is also an element in one ergodic class of **P**(2), or in general, every element in one ergodic class of **P**(n) is also an element of one ergodic class of **P**($n + 1$). Thus, by induction, the product of **P**(1), **P**(2), ..., **P**(N) includes all M^N models into a single ergodic class. Therefore, the Markov chain is irreducible.

4.2.1.3 Aperiodicity

The Markov chain with transition matrix **P** is aperiodic if for all the M^N models, $P_{ii} = p_{ii}^{(1)} > 0$, for every i. Following Rothman (1986), we will show that this is indeed true.

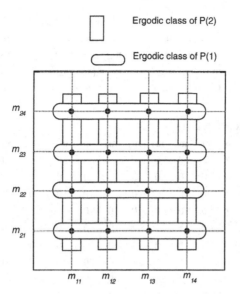

Figure 4.7 An example of a heat bath search for a model with four parameters where each parameter is allowed to have four values.

As an example, consider a model with two parameters such that the transition matrix **P** is given as

$$\mathbf{P} = \mathbf{P}(1)\mathbf{P}(2)$$
$$= \sum_{k} P_{ik}(1) P_{kj}(2), \qquad (4.21)$$

and the diagonal elements are

$$P_{ii} = \sum_{k} P_{ik}(1) P_{ki}(2). \qquad (4.22)$$

Also note the fact that each row of **P**(n) contains only M non-zero elements. This is so because for each model parameter, only M transitions are possible. This is explained in Figure 4.8a, b for $N = 2$ and $M = 4$. In this case, **P**(1) or **P**(2) contains only four non-zero elements in each row, one of which is always in the diagonal. The diagonal term is non-zero because there is always a finite non-zero probability of transition to the same state.

Therefore, $P_{ii}(1) > 0$ and $P_{ii}(2) > 0$, and all other elements are non-negative. Then

$$P_{ii} > 0. \qquad (4.23)$$

Thus the Markov chain is aperiodic.

For model parameters m_{ij}, $i=1, 2$ and $j=1, 2, 3, 4$, the possible parameters are:

$$\mathbf{m}_1 = (m_{11}, m_{21})^T \qquad \mathbf{m}_5 = (m_{11}, m_{22})^T \qquad \mathbf{m}_9 = (m_{11}, m_{23})^T \qquad \mathbf{m}_{13} = (m_{11}, m_{24})^T$$

$$\mathbf{m}_2 = (m_{12}, m_{21})^T \qquad \mathbf{m}_6 = (m_{12}, m_{22})^T \qquad \mathbf{m}_{10} = (m_{12}, m_{23})^T \qquad \mathbf{m}_{14} = (m_{11}, m_{24})^T$$

$$\mathbf{m}_3 = (m_{13}, m_{21})^T \qquad \mathbf{m}_7 = (m_{13}, m_{22})^T \qquad \mathbf{m}_{11} = (m_{13}, m_{23})^T \qquad \mathbf{m}_{15} = (m_{11}, m_{24})^T$$

$$\mathbf{m}_4 = (m_{14}, m_{21})^T \qquad \mathbf{m}_8 = (m_{14}, m_{22})^T \qquad \mathbf{m}_{12} = (m_{14}, m_{23})^T \qquad \mathbf{m}_{16} = (m_{11}, m_{24})^T$$

(a)

The Transition matrix $\mathbf{P}(1)$ is given as:

	$\mathbf{m}_1$	$\mathbf{m}_2$	$\mathbf{m}_3$	$\mathbf{m}_4$	$\mathbf{m}_5$	$\mathbf{m}_6$	$\mathbf{m}_7$	$\mathbf{m}_8$	$\mathbf{m}_9$	$\mathbf{m}_{10}$	$\mathbf{m}_{11}$	$\mathbf{m}_{12}$	$\mathbf{m}_{13}$	$\mathbf{m}_{14}$	$\mathbf{m}_{15}$	$\mathbf{m}_{16}$
$\mathbf{m}_1$	*	*	*	*	0	0	0	0	0	0	0	0	0	0	0	0
$\mathbf{m}_2$	*	*	*	*	0	0	0	0	0	0	0	0	0	0	0	0
$\mathbf{m}_3$	*	*	*	*	0	0	0	0	0	0	0	0	0	0	0	0
$\mathbf{m}_4$	*	*	*	*	0	0	0	0	0	0	0	0	0	0	0	0
$\mathbf{m}_5$	0	0	0	0	*	*	*	*	0	0	0	0	0	0	0	0
$\mathbf{m}_6$	0	0	0	0	*	*	*	*	0	0	0	0	0	0	0	0
$\mathbf{m}_7$	0	0	0	0	*	*	*	*	0	0	0	0	0	0	0	0
$\mathbf{m}_8$	0	0	0	0	*	*	*	*	0	0	0	0	0	0	0	0
$\mathbf{m}_9$	0	0	0	0	0	0	0	0	*	*	*	*	0	0	0	0
$\mathbf{m}_{10}$	0	0	0	0	0	0	0	0	*	*	*	*	0	0	0	0
$\mathbf{m}_{11}$	0	0	0	0	0	0	0	0	*	*	*	*	0	0	0	0
$\mathbf{m}_{12}$	0	0	0	0	0	0	0	0	*	*	*	*	0	0	0	0
$\mathbf{m}_{13}$	0	0	0	0	0	0	0	0	0	0	0	0	*	*	*	*
$\mathbf{m}_{14}$	0	0	0	0	0	0	0	0	0	0	0	0	*	*	*	*
$\mathbf{m}_{15}$	0	0	0	0	0	0	0	0	0	0	0	0	*	*	*	*
$\mathbf{m}_{16}$	0	0	0	0	0	0	0	0	0	0	0	0	*	*	*	*

(b)

Figure 4.8 (a) For a model vector $\mathbf{m}$ with only two model parameters such that if each can take four possible values, there are sixteen models possible. In a heat bath algorithm, all possible values of one model parameter are tested once in each iteration. (b) The transition matrix $\mathbf{P}(1)$; elements marked with an asterisk are the only non-zero values. Note that the diagonal term is non-zero because there is a finite positive probability of retaining the same model parameter value.

4.2.1.4 Limiting probability

Again we follow Rothman (1986) to show that the Gibbs distribution is indeed the limiting or stationary distribution. First, we recall that in the heat bath method current values of the neighboring model parameters determine the conditional probability distribution for any particular model parameter, and thus the process can be modeled with a Markov chain. We also showed that the Markov chain is irreducible and aperiodic.

If the symbol $P_{ij}(n)$ is used to represent the probability of transition from a model $\mathbf{m}_i$ to a model $\mathbf{m}_j$ such that only the nth model parameter is perturbed,

$$P_{ij}(n) = \frac{\exp\left[-\dfrac{E(\mathbf{m}_j)}{T}\right]}{\displaystyle\sum_{j\in A_i(n)}\exp\left[-\dfrac{E(\mathbf{m}_j)}{T}\right]}, \tag{4.24}$$

where $A_i(n)$ is the set of M indices corresponding to model parameter n. This means that model vectors $\mathbf{m}_i$ and $\mathbf{m}_j$ have the same values for all the model parameters except the nth model parameter.

Now we assume that the stationary pdf is given by the Gibbs pdf q_j such that

$$q_j = \frac{\exp\left[-\dfrac{E(\mathbf{m}_j)}{T}\right]}{\displaystyle\sum_{j\in A}\exp\left[-\dfrac{E(\mathbf{m}_j)}{T}\right]}, \tag{4.25}$$

where the set A consists of all M^N models. We now need to prove the property (Eq. 1.77); i.e.,

$$q_j = \sum_{i\in A} q_i P_{ij}. \tag{4.26}$$

First we will show that

$$q_j = \sum_{i\in A} q_i P_{ij}(n).$$

From Eq. (4.24) we know that for a given i, $P_{ij}(n)$ is non-zero only if $j \in A_i(n)$. Similarly, for a given j, $P_{ij}(n)$ is non-zero if $i \in A_j(n)$. Thus we have

$$\sum_{i\in A} q_i P_{ij}(n) = \sum_{i\in A_j(n)} q_i P_{ij}(n). \tag{4.27}$$

Now substituting Eqs. (4.24) and (4.25) into Eq. (4.27), we obtain

$$\sum_{i\in A} q_i P_{ij}(n) = \sum_{i\in A_j(n)} \left\{ \frac{\exp\left[-\dfrac{E(\mathbf{m}_i)}{T}\right]}{\displaystyle\sum_{i\in A}\exp\left[-\dfrac{E(\mathbf{m}_i)}{T}\right]} \right\} \left\{ \frac{\exp\left[-\dfrac{E(\mathbf{m}_j)}{T}\right]}{\displaystyle\sum_{j\in A_i(n)}\exp\left[-\dfrac{E(\mathbf{m}_j)}{T}\right]} \right\}$$

$$= \frac{\exp\left[-\dfrac{E(\mathbf{m}_j)}{T}\right]}{\displaystyle\sum_{i\in A}\exp\left[-\dfrac{E(\mathbf{m}_i)}{T}\right]} \frac{\displaystyle\sum_{i\in A_j(n)}\exp\left[-\dfrac{E(\mathbf{m}_i)}{T}\right]}{\displaystyle\sum_{j\in A_i(n)}\exp\left[-\dfrac{E(\mathbf{m}_j)}{T}\right]}.$$

However, when $j \in A_i(n)$, $A_j(n) = A_i(n)$. Thus we have

$$q_j = \sum_{i \in A} q_i P_{ij}(n).$$

(4.28)

This means that $\mathbf{q} = \mathbf{q}\mathbf{P}(n)$. Now substituting Eq. (4.20) into the preceding equation, we have $\mathbf{q} = \mathbf{q}\mathbf{P}$. This proves that the stationary distribution of the Markov chain in the case of the heat bath algorithm is also given by the Gibbs distribution.

4.3 Simulated annealing without rejected moves

Like the heat bath algorithm, SA without rejected moves (SAWR) (Greene and Supowit 1986) also attempts to minimize the low acceptance-to-rejection ratio of the Metropolis SA, especially at low temperatures. The method of selection, how-ever, is different from that used by the heat bath algorithm. Unlike Metropolis SA, this method determines the effects of each trial move on the energy function and uses this information to bias the selection of moves. This is done as follows:

First, we start with a random model (say, $\mathbf{m}_0$) with energy $E(\mathbf{m}_0)$. Then, for each trial move i, where $1 \leq I \leq N$, the following function is calculated:

$$P_i = \frac{w_i}{\sum\limits_{j=1} w_j},$$

(4.29)

where

$$w_i = \exp\left\{-\frac{\left[E(\mathbf{m}_i) - E(\mathbf{m}_0)\right]^+}{T}\right\}.$$

(4.30)

A model is then drawn from the preceding distribution. The procedure for doing this is exactly the same as that used in the heat bath algorithm described in Section 4.2 [Eq. (4.18), Figure 4.6]. Let us assume that the model selected is $i = k$. Now $\mathbf{m}_0$ is replaced with $\mathbf{m}_k$, a new randomly selected model is next included in the set, the weights are updated, and the process of drawing a model from the updated Eq. (4.29) is repeated. The temperature is lowered following a cooling schedule after a fixed number of iterations at each temperature, and the process is repeated until convergence.

Thus it is quite obvious that SAWR draws a model from a biased distribution, which is always updated and then retained. Since SAWR keeps track of all trial moves as it proceeds, it accumulates more global information on the objective function than the classical Metropolis SA. Also, unlike the heat bath algorithm,

SAWR does not construct the pdf by perturbing just one model parameter at a time, although a combined heat bath–SAWR algorithm may easily be formulated.

Greene and Supowit (1986) have shown that the sequence of models generated by the SAWR is equivalent to the sequence of states generated by the Metropolis method if repetition of the current model is omitted each time a move is rejected. They have also shown that in their application to a logic partitioning problem, the SAWR performs better than classical Metropolis SA at low temperatures.

4.4 Fast simulated annealing (FSA)

The Markov chain model of SA can be used to prove that the asymptotic convergence to a stationary distribution given by the Boltzmann distribution can be attained, and in the limit when the temperature goes to zero, the global minimum energy state has a probability of 1. However, whether or not the global minimum energy state can be reached depends on the rate that the temperature is lowered; i.e., it depends on the cooling schedule. Geman and Geman (1984) showed that a necessary and sufficient condition for convergence to the global minimum for SA is given by the following cooling schedule:

$$T(k) = \frac{T_0}{\ln(k)}, \tag{4.31}$$

where $T(k)$ is the temperature at iteration k and T_0 is a sufficiently high starting temperature. This logarithmic cooling schedule is very slow and therefore requires a long time to converge to the globally optimal solution.

Szu and Harley (1987) addressed this issue and proposed a new algorithm that they called *fast simulated annealing* (FSA) that is very similar to Metropolis SA except that unlike Metropolis SA, it uses a Cauchy distribution rather than a flat distribution for the model parameters to generate the models for testing using the Metropolis criterion. The Cauchy-like distribution (the generation probability) is also a function of temperature and is given by

$$f(\Delta m_i) \propto \frac{T}{\left(\Delta m_i^2 + T^2\right)^{1/2}}, \tag{4.32}$$

where T is lowered according to a specified cooling schedule, and Δm_i is the perturbation in the model parameter with respect to the current value of a model parameter.

A Cauchy distribution and its variation with temperature are shown in Figure 4.9. Note that at high temperatures, the distribution allows for selection of models

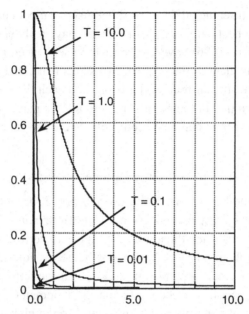

Figure 4.9 Cauchy-like distribution as a function of temperature. When this distribution is used in model generation, model perturbations are generated in the close neighborhood of the current position at low temperatures. However, at high temperatures, the current models may have substantial perturbations.

far away from the current position, whereas at low temperatures, it prefers models in the close neighborhood of the current location. The Cauchy distribution has a flatter tail than an equivalent Gaussian distribution and therefore has a better chance of getting out of local minima. Although the model-generation criterion of the FSA is different from that of Metropolis SA, the acceptance criterion is the same as that of Metropolis SA. Szu and Hartley (1987) further showed that due to the choice of a generation function given by Eq. (4.32), the cooling schedule required for convergence is no longer logarithmic, and the temperature schedule is now inversely proportional to the iteration number; i.e.,

$$T(k) = \frac{T_0}{k}. \tag{4.33}$$

The cooling schedules of Eqs. (4.31) and (4.33) are compared in Figure 4.10.

4.5 Very fast simulated reannealing

Ingber (1989, 1993) proposed several modifications of SA and proposed a new algorithm called *very fast simulated reannealing* (VFSR). The modifications were introduced for the following reasons:

Cooling Schedules

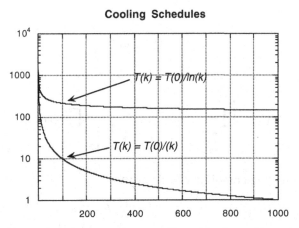

Figure 4.10 Comparison between linear and logarithmic cooling schedules.

- In an *NM*-dimensional model space, each model parameter has a different finite range of variations, and each model parameter may affect the misfit or energy function in a different manner. Therefore, they should each be allowed to have different degrees of perturbations from their current position.
- Although FSA is very elegant, there is no quick algorithm for calculating an *NM*-dimensional Cauchy random generator (Ingber 1993).

The problem of constructing an *NM*-dimensional Cauchy distribution can be avoided by using an *NM* product of 1D Cauchy distributions. In such a formulation, we can also use different temperatures, each corresponding to different model parameters. The use of a product of *NM* 1D Cauchy distributions as a generating function, however, again requires a very slow cooling schedule for convergence (Ingber 1993). Ingber therefore proposed a new probability distribution for model generation such that a slow cooling schedule is no longer required. Ingber's algorithm can be described as follows: Assume that a model parameter m_i at iteration (annealing step or time) k is represented by m_i^k such that

$$m_i^{\min} \leq m_i^k \leq m_i^{\max}, \tag{4.34}$$

where m_i^{min} and m_i^{max} are the minimum and maximum values of the model parameter m_i. This model parameter value is perturbed at iteration $k + 1$ by using the following relation

$$m_i^{k+1} = m_i^k + y_i \left(m_i^{\max} - m_i^{\min} \right) \tag{4.35}$$

such that

$$y \in \left[-1, 1 \right] \tag{4.36}$$

and

$$m_i^{\min} \leq m_i^{k+1} \leq m_i^{\max}. \tag{4.37}$$

The parameter y_i is generated from the following distribution:

$$g_T(y) = \prod_{i=1}^{NM} \frac{1}{2(|y_i| + T) \ln\left(1 + \frac{1}{T_i}\right)} = \prod_{i=1}^{NM} g_{Ti}(y_i), \tag{4.38}$$

which has the following cumulative probability:

$$G_{Ti}(y_i) = \frac{1}{2} + \frac{\operatorname{sgn}(y_i)}{2} \frac{\ln\left(1 + \frac{|y_i|}{T_i}\right)}{\ln\left(1 + \frac{1}{T_i}\right)}. \tag{4.39}$$

Thus a random number u_i drawn from a uniform distribution $U[0,1]$ can be mapped into the preceding distribution with the following formula:

$$y_i = \operatorname{sgn}\left(u_i - \frac{1}{2}\right) T_i \left[\left(1 + \frac{1}{T_i}\right)^{|2u_i - 1|} - 1\right]. \tag{4.40}$$

Ingber (1989) showed that for such a distribution, the global minimum can be statistically obtained by using the following cooling schedule:

$$T_i(k) = T_{0i} \exp\left(-c_i k^{1/NM}\right), \tag{4.41}$$

where T_{0i} is the initial temperature for model parameter i and c_i is a parameter to be used to control the temperature schedule and help to tune the algorithm for specific problems.

The acceptance rule of the algorithm is the same as that used in Metropolis SA, and the algorithm as described so far can be called *very fast simulated annealing* (VFSA). The algorithm is described schematically in Figure 4.11. The salient features of VFSA are as follows:

- It generates the perturbations for the model parameters according to the distribution given by Eq. (4.38).
- It requires a temperature for each model parameter, and these can be different for different model parameters.
- It also requires a temperature to be used in the acceptance criterion, which may be different from the model parameter temperatures.

VFSA algorithm

start at a random location with $\mathbf{m}_0$ with energy $E(\mathbf{m}_0)$

loop over temperature (T)

- loop over number of random moves/temperature
-   • loop over model parameters $i = 1, \ldots, NM$
-   •   • $u_i \in U[0,1]$
-   •   • $y_i = \mathrm{sgn}\left(u_i - \dfrac{1}{2}\right) T_i^{mod}\left[\left(1 + T_i^{mod}\right)^{|2u_i - 1|} - 1\right]$
-   •   • $m_i^{new} = m_i^{old} + y^i\left(m_i^{max} - m_i^{min}\right)$
-   •   • $m_i^{min} \le m_i^{new} \le m_i^{max}$
-   • end loop
-   • now we have a new model $\mathbf{m}^{new}$
-   • $\Delta E = E(\mathbf{m}^{new}) - E\left(\mathbf{m}_0\right)$
-   • $P = \exp\left(-\dfrac{\Delta E}{T}\right)$
-   • if $\Delta E \le 0$, then
-   •   • $\mathbf{m}^0 = \mathbf{m}^{new}$
-   •   • $E(\mathbf{m}^0) = E(\mathbf{m}^{new})$
-   • end if
-   • if $\Delta E > 0$, then
-   •   • draw a random number $r = U[0, 1]$
-   •   • if $P > r$, then
-   •   •   • $\mathbf{m}^0 = \mathbf{m}^{new}$
-   •   •   • $E(\mathbf{m}^0) = E(\mathbf{m}^{new})$
-   •   • endif
-   • endif
- end loop

end loop

Figure 4.11 A pseudo-Fortran code for a VFSA algorithm. (Adapted from Ingber 1989.)

Ingber (1989), in his VFSR algorithm, included a further modification by occasionally changing the cooling schedule for a model parameter based on how the model parameter affects the cost or energy function. This is done as follows: Let the sensitivity S_i with respect to model parameter m_i be given by

$$S_i = \frac{\partial E(\mathbf{m})}{\partial m_i}. \tag{4.42}$$

If T_{ik} is the model temperature for the model parameter i at anneal time k, this is modified according to the following rule:

$$T_{ik'} = T_{ik} \left(\frac{S_{max}}{S_i} \right),$$

(4.43)

where S_{max} is the largest value of S_i. For small values of S_i, $S_{max}/S_i > 1$ and $T_{ik'} > T_{ik}$. This causes an increase in temperature, allowing the algorithm to *stretch out* the range over which relatively insensitive parameters are being searched relative to the ranges of more sensitive parameters (Ingber 1993). Since it is possible for the temperature to increase, the term *reannealing* is employed to distinguish the algorithm from SA, where only decreasing temperatures are possible according to a predefined cooling schedule.

Ingber and Rosen (1992) compared VFSR and a GA for a suite of six test functions and found VFSR to be orders of magnitude more efficient than GA. This result may not be universally true. This is so because VFSR, like GA, has several free parameters that need to be chosen properly and may require a significant amount of user experience. However, the method itself is very general and can be easily made problem-specific. VFSA has been found to be very useful in several geophysical applications. These are described in Chapter 7.

4.6 Mean field annealing

Mean field annealing (MFA) is a variant of SA that took an entirely different approach to the formulation of a SA. The theoretical approach involves deriving a deterministic set of equations that need to be solved iteratively. This can be best understood by mapping the optimization problem into a neural network, more particularly, a Boltzmann machine or a Hopfield network. The mean field approximation is well known in statistical mechanics for allowing the stochastic behavior of large systems of electronic spin elements with complex interactions to be approximated by a deterministic set of equations. To describe the theoretical development of MFA (Peterson and Anderson 1987, 1988; Peterson and Soderberg 1989), we first need to understand such concepts as a neuron, neural networks, Hopfield network, etc.

4.6.1 Neurons and neural networks

Neural networks, also called *artificial neural systems* (ANS), are mathematical models of theorized mind and brain activity. They have proved to be useful in

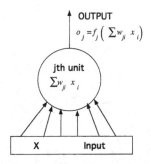

Figure 4.12 A neuron – the basic building block of a neural network. Inputs from a vector $\mathbf{x} = [x_1, x_2, x_3, \ldots, x_4]$ are multiplied with weights ω_{ij} and added. Then, by application of an activation function (usually a sigmoid), an output is obtained.

recognizing patterns and problems that can be solved by *learning through experience*. Several textbooks (e.g., Cichocki and Unbejauen 1993) describe the theory and many engineering applications of neural networks. ANS are neurally inspired models of the brain and behavior; i.e., neural models are based on how the brain might process information. The basic building block of ANS is a *neuron* or a *processing element* (PE) or a *node*, which is a multi-input, single-output element. In the brain, a massive number of neurons are operating in parallel at any given moment. Thus neural networks are dynamic systems composed of a large number of connected neurons. Processing occurs at, for example, the *j*th neuron, where a large number of signal x_i values are input. Each input signal x_i is then multiplied by a corresponding weight w_{ji} and summed over the total number of input values n (Figure 4.12) to obtain the output value y_j; i.e.,

$$y_j = \sum_{i=1}^{n} w_{ji} x_i. \tag{4.44}$$

Next, the resulting function y_j is passed through a non-linear activation function (usually a sigmoid function) to produce an output signal o_j. This function maps the total input of a neuron to an output; i.e.,

$$o_j = f(y_j) = \frac{1}{\{1 + \exp[-(y_j + \theta_j)]\}}, \tag{4.45}$$

where θ_j is the threshold that allows the sigmoid function to shift horizontally, the threshold θ_j being the amount of translation. The net effect of this shifting on a neuron is to increase or decrease the output for a given input, independently of the weights. Thus, for inputs much smaller than θ, the output equals zero, whereas for inputs on the order of θ, the neuron has an almost linear response.

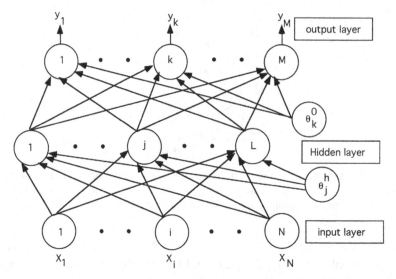

Figure 4.13 Architecture of a three-layer feed-forward partially connected net-work. In a seismic application, the input will consist of data from seismic traces, and the output will consist of model parameters.

Commonly, a one-to-one analogy is made between the artificial neuron and the biologic neuron. The inputs correspond to the *dendrites* of a biologic neuron, the weights correspond to the *synapses*, the summation and the transfer function correspond to the *cell body*, etc. Further, the firing rate of a biologic neuron is well approximated by the non-linear sigmoid function (Hopfield 1982).

In a typical neural network, neurons are arranged in layers such that two units in the same layer are not connected, and there are only connections between the neurons of successive layers. A typical example of such a three-layer, partially connected network is shown in Figure 4.13. The network consists of an input layer that holds information on the data, an output layer, and a hidden layer. The number of hidden layers is arbitrary. The optimal number of hidden layers and the number of neurons in each of the hidden layers need to be determined for specific problems.

The next step in the implementation is the training of the network. This involves computing from a pattern presented to the input layer an output pattern that is closest to the desired one. That is, we need to find a set of weights that minimize the error between the computed and desired output. The goal here is to perform a mapping from the input space, containing the input-pattern vectors as elements, to the output space, containing the output-pattern vectors. Thus the training of the network is an optimization process, which has traditionally been achieved by gradient algorithms and is called *backpropagation* (Rumelhart *et al.* 1986). The training of ANS is an area of active research, and a complete description is not of interest here. We will now describe a specific type of network called a *Hopfield network* where an optimization problem can be mapped and solved for the minimum energy state of the network.

4.6.2 Hopfield neural networks

Hopfield neural networks have significantly different architectures from that of the feed-forward neural network described in Figure 4.13. A Hopfield neural network is a single-layer feedback network whose dynamics are described by a system of non-linear ordinary differential equations and by an energy function that is minimized (Hopfield 1984; Hopfield and Tank 1985; Tank and Hopfield 1986). If $x_i(t)$, $i = 1, 2, ..., N$ is input at time t to the neurons, θ_i is the externally supplied input to neuron I, and w_{ji} is the weight connection from neuron i to neuron j, then the output at time $t + 1$, represented as $x_i(t + 1)$, is given (Cichocki and Unbejauen 1993) as

$$x_i\left(t+1\right) = \sum_{j \neq i=1}^{N} w_{ji} x_j\left(t\right) + \theta_i \begin{cases} +1 & \text{if } x_i\left(t+1\right) \geq 0, \\ -1 & \text{otherwise,} \end{cases} \tag{4.46}$$

which is essentially a step-function update rule. There are two possible ways of updating the values the neurons may take using Eq. (4.46). In each time step, if only one neuron is updated, leaving others to be computed at a future time, the network is said to be running in an *asynchronous* or *serial mode*. In a fully *parallel* or *synchronous mode* of operation, all the neurons are computed at once at every time step. This model of a Hopfield network is called a *discrete-time Hopfield model* of a neural network. For such a model, the energy function (based on physical analogy to magnetic systems) is defined as

$$E = -\frac{1}{2} \sum_{i=1}^{N} \sum_{j \neq i=1}^{N} w_{ji} x_j x_i - \sum_{i=1}^{N} x_i \theta_i, \tag{4.47}$$

where $w_{ii} = 0$. Each neuron can be in one of the two states, on (+1) or off (−1). A future state of the network can be computed from the current state using Eq. (4.46). It can be shown that the energy will never increase as the states of the neurons change if w_{ji} is symmetric.

Thus one of the most important uses of a Hopfield network is in an optimization problem in which the error or the cost function of the optimization problem is related to the one defined in Eq. (4.47) such that one can identify the connection weights, neurons, and biases. Once the minimum (usually a local minimum) energy of the network is attained, the final configuration of the neurons gives the solution of the optimization problem. Hopfield and Tank (1985) describe the traveling salesman problem in the context of the Hopfield model, and Wang and Mendel (1992) showed how a deconvolution problem can be solved using a Hopfield network by expressing the error function in such a form that the connectivity matrix, the neurons, and the biases can be identified with the problem variables. Thus the Hopfield network is distinctly different in concept from the feed-forward network described in the preceding section. In the former case, we look for the optimal weights, whereas in a Hopfield network we essentially map the model parameters into neurons, and the optimal configuration of the neurons (model parameters) corresponding to the minimum energy of the network is the solution of an optimization problem.

Of course, the first step toward achieving the minimum of the error function of the Hopfield neural network is no different from the methods described in Chapter 2; i.e., the local minimum can be obtained by gradient methods. It can be shown that a local minimum can be reached by using the relation given in Eq. (4.46) for update, assuming that the interconnection matrix w_{ij} is symmetric with non-negative elements in the main diagonal and that the neurons are updated asynchronously (e.g., Cichocki and Unbejauen 1993).

4.6.3 Avoiding local minimum: SA

The step-function update rule given by Eq. (4.46) only guarantees convergence to the nearest local minimum of the energy function. This can be avoided by using SA instead of Eq. (4.46), as is done in Boltzmann machines (e.g., Hinton *et al.* 1984) that use the Hopfield energy function and SA rules for update. That is, whether at any future time step a neuron will be on (+1) or off (−1) is not deterministic but rather is given by the following probability (as in SA):

$$P\left(x_i\left(t+1\right)=\pm1\right)=\frac{\exp\left(\pm u_i/T\right)}{\exp\left(u_i/T\right)+\exp\left(-u_i/T\right)},\tag{4.48}$$

where

$$u_i = \sum_{j=1}^{n} w_{ij}x_j + \theta_i,\tag{4.49}$$

and x_j is +1 or −1 for all j, and T is the familiar control parameter *temperature*. Thus whether x_i at step $t + 1$ will be +1 or −1 is determined by the Metropolis criterion by drawing a random number from a uniform distribution and comparing it with P, as described earlier.

4.6.4 Mean field theory (MFT)

In SA, either one or the other of the two states $(+1$ or $-1)$ is determined stochastic-ally. The idea in *mean field theory* (MFT) is to replace the stochastic neurons with continuous outputs constrained between $+1$ and -1. This is achieved by using a trick to evaluate the partition function. First, consider that we have a set of n neurons such that each neuron x_i is allowed to assume values of $+1$ and -1, and assume that the vector $\mathbf{x}$ represents a particular configuration of neurons or a state with an energy $E(\mathbf{x})$. Recall that when using SA, the probability that a network will be in a state $\mathbf{x}$ is given by the following Gibbs pdf:

$$p(\mathbf{x}) = \frac{\exp\left[-\dfrac{E(\mathbf{x})}{T}\right]}{\displaystyle\sum_{\mathbf{x}} \exp\left[-\dfrac{E(\mathbf{x})}{T}\right]}, \tag{4.50}$$

where the term in the denominator is the partition function, which requires sum-ming over all possible neuron configurations. This can be evaluated only if the energy for all possible configurations is evaluated – something we always want to avoid. However, we described in earlier sections how one can asymptotically obtain the Gibbs pdf by Markov chains. Peterson and Anderson (1987, 1988) use a simple trick to evaluate the partition function in Eq. (4.50) which results in the so-called MFT equations. We will closely follow Peterson and Anderson (1987, 1988) to describe the development of the MFT equations.

Recognizing the fact that each neuron can assume values of $+1$ or -1, the sum in the denominator of Eq. (4.50) can be expressed as the following nested sums:

$$Z = \sum_{\mathbf{x}} \exp\left[-\frac{E(\mathbf{x})}{T}\right] = \sum_{x_1=\pm 1} \cdots \sum_{x_i=\pm 1} \cdots \sum_{x_n=\pm 1} \sum_{x_n=\pm 1} \exp\left[-\frac{E(\mathbf{x})}{T}\right]. \tag{4.51}$$

Now each discrete sum can be expressed in terms of integration over a continuous variable V as

$$\sum_{x_i=\pm 1} \exp\left[-\frac{E(\mathbf{x})}{T}\right] = \sum_{x_i=\pm 1} \int_{-\infty}^{\infty} dV \, \exp\left[-\frac{E(V)}{T}\right] \delta(\mathbf{x} - V) \tag{4.52}$$

by using the property of the delta function. The delta function can also be repre-sented by the following integral:

$$\delta(V) = \frac{1}{2\pi i} \int_{-i\infty}^{i\infty} dU \, \exp(UV). \tag{4.53}$$

Substituting Eq. (4.53) into Eq. (4.52), we obtain

$$
\sum_{x_i=\pm 1} \exp\left[-\frac{E(\mathbf{x})}{T}\right] = \frac{1}{2\pi i} \sum_{x_i=\pm 1} \int_{-\infty}^{\infty} dV \int_{-i\infty}^{i\infty} dU \exp\left[-\frac{E(V)}{T}\right] \exp\left[U(x_i - v)\right]
$$

$$
= \frac{1}{2\pi i} \int_{-\infty}^{\infty} dV \int_{-i\infty}^{i\infty} dU \exp\left[-\frac{E(V)}{T}\right] \exp(-UV) \sum_{x_i=\pm 1} \exp(Ux_i)
$$

$$
= \frac{1}{\pi i} \int_{-\infty}^{\infty} dV \int_{-i\infty}^{i\infty} dU \exp\left[-\frac{E(V)}{T}\right] \exp\left[-UV \ \log(\cosh U)\right].
$$

(4.54)

From Eqs. (4.51), (4.52), and (4.54), and rearranging terms, we obtain

$$
Z = c \prod_i \int_{-\infty}^{\infty} \int_{-i\infty}^{i\infty} \exp\left[-E'(\mathbf{v}, \mathbf{u}, T)\right] dU_i dV_j,
$$

(4.55)

where c is a constant, and the new function E' is given by

$$
E'(\mathbf{v}, \mathbf{u}, T) = \frac{E(\mathbf{v})}{T} + \sum_j \left[U_j V_j - \log(\cosh U_j)\right].
$$

(4.56)

Notice now that the partition function, which was a function of variables x_i (+1 or −1), is now expressed in terms of the variables U_i and V_i, which are continuous variables that can take any value between +1 and −1. One important property of the new energy-like term E' is that this function is smoother than E due to the second term in Eq. (4.56) (Peterson and Anderson 1987), and thus we may attempt to evaluate Eq. (4.55) by the saddle-point method. The saddle points of Z are determined by setting the derivative of the phase term E' (Eq. 4.56) with respect to the variables U_i and V_i to zero to obtain

$$
\frac{\partial E'}{\partial U_i} = V_i - \tanh U_i = 0,
$$

(4.57)

and

$$
\frac{\partial E'}{\partial V_i} = \frac{1}{T}\frac{\partial E}{\partial V_i} + U_i = 0,
$$

(4.58)

which can be rewritten as

$$
V_i = \tanh U_i,
$$

(4.59a)

and

$$U_i = -\frac{1}{T}\frac{\partial E(\mathbf{v})}{\partial V_i}. \tag{4.59b}$$

These are called the *mean field equations*. Note that in the derivation of Eq. (4.59), no assumption was made on the architecture of the network, and the derivation of Eq. (4.59) does not depend on the functional form of the energy function.

Recall that for the particular case of a Hopfield network, the energy function for a given neuron configuration $\mathbf{v}$ can be computed by the equation

$$E(\mathbf{v}) = -\frac{1}{2}\sum_{i=1}^{n}\sum_{j=1}^{n}w_{ij}V_iV_j + \sum_{i=1}^{n}V_i\theta_i.$$

The second term can be dropped by using an extra neuron V_0 permanently in a +1 state with $w_{0i} = w_{i0} = -\theta_i$ such that

$$E(\mathbf{v}) = -\frac{1}{2}\sum_{i=0}^{n}\sum_{j=0}^{n}w_{ij}V_iV_j. \tag{4.60}$$

For a given configuration $\mathbf{v} = \{V_i \text{ (old)}\}$, E can be computed from Eq. (4.60), and $E(\mathbf{m}_j)\partial E/\partial V_i$ can also be calculated, which can then be used to compute U_i. Equation (4.59a) can now be used to compute the updated value of V_i. Using Eq. (4.60), it reduces to the following simple form:

$$V_i(\text{new}) = \tanh\left[\frac{\sum_j w_{ij}V_j(\text{old})}{T}\right]. \tag{4.61}$$

For any other energy function, the update has to be computed numerically. Equation (4.61) is a sigmoid update function, and we note that at $T = 0$ it reduces to a step function – the standard Hopfield network update formula. For high values of T, the updates are very small and nearly uniform. This is described in Figure 4.14.

Equation (4.61) can be used either in a serial or a parallel mode. Peterson and Anderson (1987) noted that the serial method (or asynchronous update) appears to be advantageous.

So far we have closely followed Peterson and Anderson's (1987) derivation of the MFT equations – a deterministic set of equations starting from the familiar SA partition function, which can be used to compute a new state (configuration) from a current state that includes temperature dependence. But unlike classical Metropolis SA or any variant thereof, such as VFSA, there is no stochastic acceptance/rejection criterion. In the derivation of the MFT equations, the original variables x_i

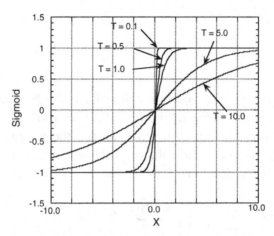

Figure 4.14 Tangent hyperbolic functions for different temperatures (Eq. 4.61).
Note that at low temperature the function reduces to a step-function update rule.

were replaced with V_i, but so far we have not discussed what the V_i represent.
Peterson and Anderson (1987) showed, by using a Markov model (see the appendix of Peterson and Anderson, 1987), that the variables V_i are the mean (or average) values of the variables x_i and are given by

$$V_i = \langle x_i \rangle \big|_T = \frac{x_i(=+1)P(x_i=+1) + x_i(=-1)P(x_i=-1)}{P(x_i=+1) + P(x_i=-1)}$$

$$= P(x_i=+1) - P(x_i=-1).$$

(4.62)

Now that we understand the meaning of the variables V_i and we have the MFT update rule (once the energy function has been identified), we are now in a position to describe an algorithm to solve an optimization problem. But how do all these ideas relate to our geophysical inversion problem? We mentioned earlier that the model parameters may be identified with neurons, but each one of our *NM* model parameters is allowed to take a finite number of values. For example, consider the example in Figure 4.6, in which we have eight model parameters, each of which can have eight possible values. To describe a complete trial model, each model parameter can have only one of its eight possible values. Given that each neuron can be either on (+1) or off (−1), how do we represent this? This problem can be solved by introducing the multistate form based on what is known as a *Potts glass model* (Peterson and Soderberg 1989). In such a scheme, rather than using x_i, the neurons are arranged in nodes, and a second index is introduced. For example, now we will use x_{ik} to represent the *k*th value of the *i*th model parameter, or *i* denotes the node and *k* denotes the neuron within each node such that

$$\sum_k x_{ik} = 1. \tag{4.63}$$

Also in the Potts glass model, the two possible states of each neuron are now changed to 0, 1 rather than −1, 1. Equation (4.63) ensures that each model parameter takes only one value. Thus the variables in the MFT equations described earlier have an extra dimension added to each one of them. Each x_{ik} is like an element of a matrix such that a vector x_i has k elements and each element is either 0 or 1. Thus the equivalent of the mean field variable V_i is now a vector v_i such that

$$\mathbf{v}_i = \langle \mathbf{x}_i \rangle |_T, \tag{4.64}$$

each component of which can be calculated using an equation similar to Eq. (4.62). The MFT equations now take the following form:

$$\mathbf{v}_i = \mathbf{F}_k(\mathbf{u}_i)$$
$$\mathbf{u}_i = -\frac{1}{T}\frac{\partial E}{\partial \mathbf{V}_i}, \tag{4.65}$$

where each component of V_{ij} is written as

$$v_{ij} = \frac{\exp(U_{ij})}{\sum_k \exp(U_{ik})}. \tag{4.66}$$

Clearly,

$$\sum_j V_{ij} = 1. \tag{4.67}$$

Thus, for a given configuration V_i, the energy E, its derivative with respect to each $\mathbf{v}_i$ and $\mathbf{u}_i$, can be computed. The $\mathbf{u}_i$ can now be used to compute update values of $\mathbf{v}_i$ either in synchronous or asynchronous mode. Also recall that each V_{ij} is the mean value of x_{ij} and takes any value between 0 and 1. Thus V_{ij} can be interpreted as the probability that an $x_{ij} = 1$. Thus, unlike all other SA algorithms described earlier, *MFA works with the probabilities of the variables rather than the variables themselves.*

MFA can be initiated with nearly uniform values of the probability variables, and Eq. (4.65) can be used iteratively (either in synchronous or asynchronous mode) while lowering the temperature using a cooling schedule. The final result will give the highest probabilities for each of the model parameters, i.e., the answer to the problem.

Although this appears like a very simple algorithm, the choice of the energy function is critical in that it should not be computationally expensive. Of course, the first requirement is that the energy function be a function of probability variables. One way to do this is to write the energy function in a form very similar to that used in a Hopfield network, as has been done by Wang and Mendel (1992). This, however, is not a trivial task for every problem. Upham and Cary (personal communication, 1993) described an error function simply as a quadratic function of probability variables.

One potential advantage of MFA is that an approximate formula for the critical temperature has been derived by Peterson and Soderberg (1989). The MFA can be started near or slightly above this critical temperature such that we can reach near to the global minimum of the energy function very rapidly. However, the computation of the critical temperature by the method of Peterson and Soderberg (1989) is also not trivial because it requires calculating the eigenvalues of a Hessian matrix (second derivative of error with respect to the probability variables), which can be large in size for geophysical applications.

4.7 Using SA in geophysical inversion

We have described several variants of SA methods that attempt to sample model space according to a Gibbs distribution that depends on a control parameter called *temperature*. Thus, if used properly, the algorithm can find the global minimum of the error function without the requirement of starting close to it. But to describe the solution of the inverse problem, we also need to know the uncertainties in the final result. At the minimum, we can compute the covariance in $\mathbf{m}_{est}$ by evaluating the second derivative of the error with respect to model parameters evaluated at $\mathbf{m}_{est}$ as given in Eq. (2.48). This would, however, assume a Gaussian posterior with mean given by $\mathbf{m}_{est}$. This, in itself, would be a major improvement over gradient methods in the absence of good starting models. However, we can also do better than this with SA, as we describe below.

4.7.1 Bayesian formulation

SA can also be described using a Bayesian approach, as shown by Geman and Geman (1984) and Rothman (1985). They showed using Bayes' rule that the posterior pdf is given by the Gibbs pdf, and the noise need not be Gaussian. Thus, locating the global minimum of an error function corresponds to searching the maximum of the posterior probability density function (PPD), and therefore an SA with cooling is a maximum a posteriori (MAP) estimation algorithm.

Recall that most SA algorithms are statistically guaranteed to draw samples from a stationary distribution at a constant temperature, which is the Gibbs distribution. Thus, if one repeats the model-generation acceptance procedure of SA at constant temperature a large number of times, the models are distributed according to the Gibbs distribution.

Tarantola (1987) made an interesting observation about this point. Recall that the PPD is given by

$$\sigma\left(\mathbf{m}|\mathbf{d}_{obs}\right) \propto \exp\left[-E(\mathbf{m})\right], \tag{4.68}$$

which is the same as the Gibbs distribution when we set $T = 1$. Thus, if we generate several models by SA at a constant temperature of $T = 1$, the models are distributed according to the PPD. Thus the SA technique (at a constant temperature) can be used in the evaluation of multidimensional integrals (Eq. 2.79) by importance sampling techniques. We will describe this concept with some examples in detail in Chapter 8.

Basu and Frazer (1990) proposed that the samples be drawn at the critical temperature and detail a numerical approach to estimating the critical temperature. We note, however, that the critical temperature can only be estimated numerically at a substantial computational cost (Basu and Frazer 1990; Sen and Stoffa 1991). Also, there is no theoretical basis to assume that drawing only a few models will be sufficient to obtain an equilibrium distribution at critical temperature and obtain unbiased estimates of the integral (Eq. 2.79). Also at critical temperature, we sample models from a distribution that is proportional to $\exp[-E(m)/T_c]$, where T_c is the critical temperature. Usually T_c is a number smaller than 1. Thus the distribution $\exp[-E(m)/T_c]$ is much narrower than the PPD (Eq. 4.68). Therefore, the posterior model parameter variances will be underestimated.

4.8 Summary

Since the work of Kirkpatrick *et al.* (1983), the optimization method now known as SA has been applied to several problems in science and engineering. Following the work of Rothman (1985, 1986), the method has gained popularity in the geophysical community and has been applied to a wide variety of problems in geophysics. Many geophysical inverse problems are non-linear, and due to the availability of high-speed computers, SA is now being applied to these problems. In this chapter we reviewed, starting with classical SA, several variants of SA that are aimed at making the algorithm computationally efficient. Most of these algorithms are statistically guaranteed to attain equilibrium distribution and possibly reach the global minimum of an error or energy function and are thus suitable in MAP estimation. We also note that they can be used as importance sampling algorithms to estimate the uncertainties and the marginal posterior probability density function for each parameter in model space. In practice, however, these algorithms cannot be expected to always perform flawlessly because care needs to be taken in parameter selection, which is problem-dependent. For example, it is important to choose the starting temperature and cooling schedule properly. One obvious omission in this chapter is a discussion of cooling schedules. There has been some theoretical work on choosing a cooling schedule. However, the problem with theoretical cooling schedules is that your computer budget will quickly be exceeded. For this reason, we found VFSA to be very useful in many applications. The SAs that use fast cooling schedules are often called *simulated quenching* (SQ) (Ingber 1993). One variant of SQ is to find the critical temperature and either run the algorithm at that constant temperature (Basu and Frazer 1990) or anneal slowly by starting slightly above the critical temperature (Sen and Stoffa 1991). However, a substantial amount of computer time can be spent finding the critical temperature. The MFA algorithm appears promising in that an approximate value of the critical temperature can be found analytically. SA at a constant temperature of one avoids the issues of selecting a suitable starting temperature and a cooling schedule and enables us to use SA as an importance sampling technique in the efficient evaluation of multidimensional integrals. These concepts are illustrated in Chapter 8. In MAP estimations, however, choice of the proper cooling schedule is critical to the design of an efficient algorithm. In practice, an SQ method that uses a cooling schedule that is faster than a theoretical cooling schedule can be determined by trial and error.

5

Genetic algorithms

Unlike simulated annealing (SA), which is based on analogy with a physical annealing process, genetic algorithms (GAs), first proposed by John Holland (1975), are based on analogies with the processes of biologic evolution. Thus they can also be called *simulated evolution* (Fogel 1991). Classical GAs, as proposed by Holland (1975) and described in Goldberg (1989), differ significantly from classical SA (Kirkpatrick *et al.* 1983). Unlike SA, an initial population of models is selected at random, and the GA seeks to improve the fitness (which is a measure of goodness-of-fit between data and synthetics for the model) of the population generation after generation. This is principally accomplished by the genetic processes of selection, crossover, and mutation. Excellent reviews of GAs are given in the books by Goldberg (1989) and Davis (1991) and the article by Forrest (1993). Here we present a brief review of a *classical* GA (also called a *simple GA* in Goldberg 1989) and then describe some of the modifications to the approach (e.g., Stoffa and Sen 1991; Sen and Stoffa 1992a) that were found beneficial in geophysical applications. In particular, we describe how some of the concepts employed in SA can be incorporated into GAs. This has at least a conceptual advantage in describing and relating these two non-linear optimization procedures. We also believe that they introduce a performance advantage.

5.1 A classical GA

Unlike SA, a GA works with a population of models, and also unlike any other optimization method, it works with models that are coded in some suitable form. The basic steps in a GA are coding, selection, crossover, and mutation.

5.1.1 Coding

Common to any basic GA is the discretization of the list of model parameters using a binary coding scheme that results in an analogue of the chromosome.

Higher-order coding schemes are possible, as are more complex representations, such as diploid and multiple chromosomes (see, e.g., Davis and Coombs 1987; Glover 1987; Koza 1992; Smith and Goldberg 1992).

The initial step in employing a GA is to design a coding scheme that represents the possible solutions to a given problem. In geophysical applications, this is usually straightforward. We assume that we know the physical parameters that control the observed geophysical response we have measured. Each of these physical parameters becomes a model parameter that will be searched in combination with all the other model parameters to find a solution that best agrees with the observed geophysical data.

In the simple binary coding scheme, each bit corresponds to a *gene* that can take a value of 0 or 1 called an *allele*, and each individual in the population is completely described by its bit string or *chromosome*. This genetic information will be acted on by the algorithm in such a way as to favor models whose list of model parameters produces a fit response that describes the observed data very well.

This gives rise to the second step when seeking to employ a GA: the definition of the fitness. This is also problem-dependent, but in geophysical optimization problems we can employ variations in the usual norms, L_2, L_1, etc., such that goodness-of-fit (called the *fitness function*) instead of misfit is defined as the objective function to be maximized.

In a GA, the coding scheme for each model parameter can be different. This means that the search space of each model parameter can be independently defined and its resolution independently specified. Thus the basic concept of coding the physical parameters using a binary representation limits the search space, defines the resolution of each model parameter, and limits the number of acceptable models.

Consider a problem where different geologic units have different physical properties, and these properties represent the model parameters to be estimated. Consider, for example, the coding of compressional-wave velocity. For one geologic unit, the low-velocity limit of interest may be 1,500 m/s and the upper limit 1,810 m/s. Assume that the desired resolution is 10 m/s. For this coding, 5 bits are required (Figure 5.1). When all the bits are 0, the value of 1,500 m/s is represented. If the first bit becomes 1, the value of 1,510 m/s is obtained. If the first and second bits are 1, the compressional-wave velocity is 1,530 m/s. If all five bits are 1, the velocity represented is 1,810 m/s.

The compressional-wave velocity for another geologic unit is another model parameter, and its coding may be different. For example, when all the bits are 0, the minimum may now correspond to 2,000 m/s, and the resolution may be reduced to 20 m/s. If the maximum velocity of interest is now 4,540 m/s, 7 bits are required for this model parameter.

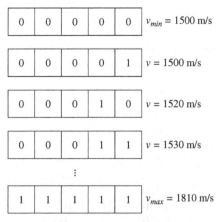

Figure 5.1 An example of binary coding of model parameters as used in a GA. Assuming that the seismic wave velocity of a rock layer varies between 1,500 and 1,810 m/s at a resolution of 10 m/s, we can represent all possible values within the range using 5 bits. When all the bits are off, we have a minimum value of the model parameter. Similarly, when all the bits are on, we have a maximum value of the model parameter. Such binary-coded model parameters are concatenated one after another to form a *chromosome* or a model.

One problem with simple binary coding is that we may need to change several bits to increment a model parameter value by one unit. For example, consider the velocity model in Figure 5.1, in which v_{min} = 1,500 m/s, Δv = 10 m/s, and v_{max} = 1,810 m/s, and examine the binary representations of the two following velocity values:

00111 = 1,570 m/s
01000 = 1,580 m/s

Notice that in order to change the velocity by 10 m/s, we must alter 4 bits. Gray coding (Forrest 1993), logarithmic scaling, and delta coding (Whitley *et al.* 1991) have been used to avoid some of the problems associated with simple binary coding. A Gray code is such that the binary representation of two consecutive Gray-coded numbers differs by only 1 bit, as shown in Table 5.1. Gray codes are generated by forming the *bitwise exclusive or* of an integer i with the integer part of $i/2$ (Press *et al.* 1989).

Once a coding scheme has been selected, all the coded model parameters are joined together into a long bit string, analogous to a *chromosome*, which now represents the genetic information of each individual that will be modified by the algorithm. The length of this string in bits has often been used as a guide to the number of individual models that need to be in the population (Goldberg 1989; Berg 1990). Typically, the individual models are initially chosen at random from

Table 5.1. *Binary codes and Gray-coded binary representation of some integers*

Integer	Gray code	Binary code
0	0000	0000
1	0001	0001
2	0011	0010
3	0010	0011
4	0110	0100
5	0111	0101
6	0101	0110
7	0100	0111
8	1100	1000
9	1101	1001
10	1111	1010
11	1110	1011
12	1010	1100
13	1011	1101
14	1001	1110
15	1000	1111

the discrete model space. The algorithm must now determine the fitness of the individual models. This means that the binary information is decoded into the physical model parameters, and the forward problem is solved. The resulting synthetic data estimate is then compared with the actual observed data using the specified fitness criterion. Depending on the problem, the definition of fitness will vary. For example, two normalized correlation functions (Sen and Stoffa 1991) recently used in geophysical inversion are given below:

$$F(\mathbf{m}) = \frac{u_o \otimes u_s(\mathbf{m})}{\left(u_o \otimes u_s\right)^{1/2} u_s(\mathbf{m}) \otimes u_s(\mathbf{m})^{1/2}},$$ (5.1)

and

$$F(\mathbf{m}) = \frac{2u_o \otimes u_s(\mathbf{m})}{\left(u_o \otimes u_o\right) + \left(u_s(\mathbf{m}) \otimes u_s(\mathbf{m})\right)},$$ (5.2)

where u_o and $u_s(\mathbf{m})$ correspond to the observed data and synthetic data for model $\mathbf{m}$, respectively, and $\otimes$ represents correlation. Note that the fitness $F(\mathbf{m})$ is bounded between -1 and $+1$. One may also define fitness as

$$F_1(\mathbf{m}) = 0.5\left(1 + F(\mathbf{m})\right),$$ (5.3)

which is always positive and bounded between 0 and 1. One may also use the negative of an error function (e.g., as defined in Chapter 2) as the fitness function.

5.1.2 Selection

Once the fitness of each individual model in the population is determined, the first genetic process, selection, is employed. Selection pairs individual models for reproduction based on their fitness values. Models with higher fitness values are more likely to get selected than models with lower fitness values. This is the analogue of inheritance because fitter models will be selected for reproduction more often. The probability of selection can be directly related to the fitness value or some function of the fitness (see Goldberg 1989 for details). It is also possible to arbitrarily eliminate or select models that are below or above specified fitness thresholds. There exist at least three methods of selection: (1) fitness-proportionate selection; (2) rank selection, and (3) tournament selection.

5.1.2.1 Fitness-proportionate selection

The most basic selection method, *stochastic sampling*, uses the ratio of each model's fitness function to the sum of fitness values of all the models in the population to define its probability of selection; i.e.,

$$p_s\left(\mathbf{m}_i\right) = \frac{F\left(\mathbf{m}_i\right)}{\sum_{j=1}^{n} F\left(\mathbf{m}_j\right)}, \tag{5.4}$$

where n is the number of models in the population. Selection based on these probabilities proceeds until a subset of the original models has been paired. The composition of the new population will consist of the original models replicated in direct proportion to their probability multiplied by the total number of models in the population.

However, an alternative to this approach, called *stochastic remainder selection without replacement*, is usually employed. In this procedure, the expected count of each model in the new population is determined as described earlier. The integer and fractional remainder of the expected count are then determined. Each original model is replicated in direct proportion to its integer parts, and the fractional part is used as an additional probability for selection (Goldberg 1989).

5.1.2.2 Rank selection

In the rank selection method (Baker 1987; Whitley 1989), the fitness values of the models are evaluated, and then they are sorted such that each model is assigned a rank. The rank of a model may range from 0 (for the best model) to $n - 1$ (for the worst model). The selection is done in such a way that the best model in the population contributes a number of copies that is an integral (the number is predetermined) number of copies that the worst model receives. Thus rank selection essentially exaggerates the difference between models with nearly identical fitness values.

5.1.2.3 Tournament selection

This method (e.g., Goldberg and Deb 1991) simulates the competition that exists in nature among individuals for the right to mate. In the simplest version of the method, random pairs are selected from a population of, say, n individuals and their fitness values are computed. The fitness is compared, and one of the two models is accepted based on a probability P_s, called the *tournament selection number*, which is an adjustable parameter. For $P_s > 0.5$, better models are favored. This procedure is repeated until there are n models in the offspring population. Thus tournament selection is essentially a probabilistic version of rank selection. In a more general version of the method, selection can be made from a group larger than two as done in evolutionary programming (EP) (Fogel 1991), described later in this chapter.

In a basic GA, if the population originally contained 100 models, 50 pairs are selected based on their fitness values. Each pair of models will now produce two offspring using the genetic operators of crossover and mutation. This will result in a completely new population of individuals.

Alternatively, we can reject some percentage of the least-fit models, and the paired models may produce only one offspring, requiring more mating occurrences from the pool of acceptable models. Other variations include reseeding the population with additional randomly selected models to replace the least-fit models, defining a measure of distance between the multidimensional models, and using this to delete some percentage of the models that are close to one another, replacing these with models from more distant parts of the model space. In the most basic GA, the models are selected using, for example, remainder-rule selection; they are paired and produce two offsprings, and the new models completely replace the preceding generation.

5.1.3 Crossover

Once the models are selected and paired, the genetic operator of recombination, crossover, occurs. Crossover is the mechanism that allows genetic information between the paired models to be shared. In the terminology of geophysical inversion, crossover causes the exchange of some information between the paired models, thereby generating new pairs of models. Crossover can be performed in two

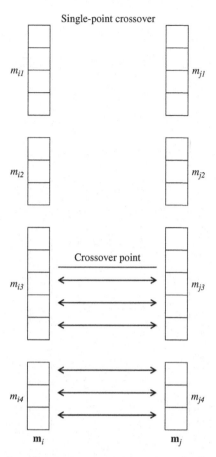

Figure 5.2 In a single-point crossover, a crossover site is selected at random between two parent chromosomes, and the bits to the right (*below*) of the crossover site are exchanged. For geophysical applications, this would mean interchanging of several model parameters between the two models.

modes: single point and multipoint. In a single-point crossover, one bit position in the binary string is selected at random from a uniform distribution. All the bits to the right of this bit are now exchanged between the two models, generating two new models. In multipoint crossover, this operation is carried out independently for each model parameter in the string. That is, for the first model parameter, a bit position is randomly selected, and the bits to the right, i.e., the lower-order bits, are exchanged between the paired models. For the second model parameter, another bit position is selected at random, and the bits to the right are again exchanged. This process is repeated for each model parameter that is concatenated into the binary string.

Figures 5.2 and 5.3 show how single- and multipoint crossover are performed. In Figure 5.2, all the bits between model $\mathbf{m}_i$ and model $\mathbf{m}_j$ below the crossover

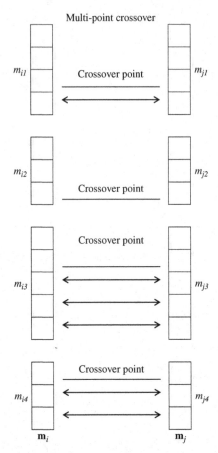

Figure 5.3 In a multipoint crossover, a crossover point is selected between each model parameter between two chromosomes. For geophysical applications, this will result in perturbing each model parameter of the two models.

point, which is selected at random, are exchanged, resulting in two new models. Multipoint crossover (Figure 5.3) exchanges binary information between each model parameter by selecting a crossover point independently within each coded model parameter for all the model parameters and performing crossover on a model-parameter-by-model-parameter basis.

Figures 5.4 and 5.5 illustrate the process of crossover for two cases. In Figure 5.4, the paired velocity model parameters v_1 and v_2 are the extreme members for the coding scheme of Figure 5.1, 1,500 m/s and 1,810 m/s. For each possible crossover point, the resulting velocity values for the new model parameters, v_1 and v_2, are shown. In a multipoint crossover scheme, if the crossover position is 5, the resulting models will have completely exchanged this particular model parameter value. For crossover positions 1, 2, 3, and 4, an interpolation has occurred. If the crossover position is 0, no change occurs. In this example, the effect of crossover

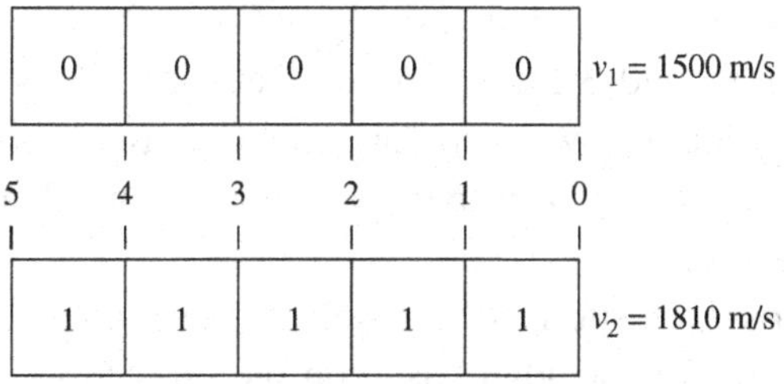

| 0 | 0 | 0 | 0 | 0 | $v_1 = 1500$ m/s |

| 5 | 4 | 3 | 2 | 1 | 0 |

| 1 | 1 | 1 | 1 | 1 | $v_2 = 1810$ m/s |

Crossover position	v_1	v_2
	(m/s)	(m/s)
0	1500	1810
1	1510	1800
2	1530	1780
3	1570	1740
4	1650	1660
5	1810	1500

Crossover Example 1

Figure 5.4 An example of crossover between two model parameters. Model parameters are the same as those shown in Figure 5.1. Note that this crossover operation results in interpolation between the two end members.

Crossover position	v_1	v_2
	(m/s)	(m/s)
0	1760	1730
1	1770	1720
2	1770	1720
3	1810	1680
4	1730	1760
5	1730	1760

Crossover Example 2

Figure 5.5 Crossover between two model parameters with values different from those shown in Figure 5.4. Model parameter limits, resolution, and coding are the same as those shown in Figure 5.1. Note that – depending on the crossover point – we can have any model parameter value within the limits with the given resolution.

is an interpolation between two extreme values. Multipoint crossover would be analogous to a multidimensional interpolation scheme.

However, crossover does not always result in interpolation, as shown in Figure 5.5. Here the values for v_1 and v_2 are closer, 1,760 and 1,730, respectively. Again, crossover at position 0 results in no change, and crossover at position 5 exchanges the velocity values exactly between the two models. For v_1, crossover at positions 1, 2, and 3 results in new velocity values greater than either of the original two velocities, whereas for velocity 2, the new velocity values are lower. Thus the effects of crossover in this example are no change, a complete exchange, or an extrapolation.

Once a crossover site has been selected at random, whether or not crossover will be performed is decided based on the probability of crossover; i.e., the rate of crossover is controlled by a probability p_x specified by the designer of the GA. A high probability of crossover means that it is likely that crossover will occur between the paired models or, in the case of multipoint crossover, between the current model parameter of the paired models.

5.1.4 Mutation

The last genetic operator is mutation. Mutation is the random alteration of a bit. It can be carried out during the crossover process. The mutation rate is also specified by a probability determined by the algorithm designer. A low mutation probability will limit the number of random walks in model space taken by the algorithm. A high mutation probability will result in a large number of random walks but may delay convergence of the algorithm to the optimally fit model. Figure 5.6 illustrates the process of mutation for an initial velocity value of 1,760 m/s. Depending on which bit is altered, the resulting velocity values are indicated. It is clear that for this simple binary coding scheme, altering a single high-order bit, e.g., bit 4 or bit 5 in Figure 5.6, has much greater impact than altering a lower-order bit.

5.2 Schemata and the fundamental theorem of genetic algorithms

Holland (1975) established a theoretical basis for GAs. A GA searches for the similarity between the models (strings) being compared and uses this information to improve the models so as to achieve better results in subsequent generations. Holland (1975) used the concept of schema or schemata. A *schema* is a template or a pattern-matching device generated by the trinary alphabet {0, 1, *}, where * is a *don't care* symbol. An example of a schema is

$$H = 0\ 1\ 1 * 1 * *. \tag{5.5}$$

The order of a schema [represented as $o(H)$] is the number of fixed positions (number of 1s and 0s) in the template. For the preceding example, $o(H) = 4$. The length of the schema [represented as $\delta(H)$] is the distance between the first and the last specific string position. For the preceding example, $\delta(H) = 4$.

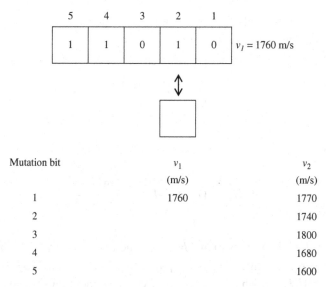

Mutation bit	v_1 (m/s)	v_2 (m/s)
1	1760	1770
2		1740
3		1800
4		1680
5		1600

Figure 5.6 An example of mutation. A mutation point is selected at random, and its bit value is altered based on a certain finite mutation probability. This operation helps to introduce diversity in the population. The top panel shows the binary code of a model parameter ($v_1 = 1{,}760$ m/s). The table below shows the velocity value resulting from mutation at different bit positions.

Implicit in the GA search is the processing of a large number of schemata generated by the trinary alphabet. Holland studied the effects of selection, crossover, and mutation on the growth and decay of schemata. He showed, by placing some limits on the probability, that short, low-order, above-average (i.e., average fitness of the models containing the schema is greater than the average fitness of the population) schema receive exponentially increasing trials in subsequent generations. This is called the *schema theorem* or the *fundamental theorem* of GAs. Goldberg (1989) describes the theorem in detail. We will summarize the theorem below based on the description given in Goldberg (1989, pp. 28–33).

Let H represent a particular schema and at a given time step (generation) t. Thus there are $M(H, t)$ examples of a particular schema present in the population of size n. Recall that during the remainder selection process, each model $\mathbf{m}_i$ in a population is reproduced according to its fitness $F(\mathbf{m}_i)$, as given by the selection probability in Eq. (5.4). Thus $M(H, t + 1)$, which represents examples of schema H at generation $(t + 1)$, can be related to $M(H, t)$ as follows:

$$M(H, t+1) = M(H, t) \frac{\bar{F}[\mathbf{m}(H)]}{\frac{1}{n}\sum_{j=1}^{n} F(\mathbf{m}_j)}, \tag{5.6}$$

where $\bar{F}[\mathbf{m}(H)]$ represents average fitness of models with schema H. Also, the average fitness of the entire population is given by

$$\langle \bar{F} \rangle = \frac{\sum\limits_{j=1}^{n} F(\mathbf{m}_j)}{n}. \tag{5.7}$$

Thus we have

$$M(H, t+1) = M(H, t) \frac{\bar{F}[\mathbf{m}(H)]}{\langle \bar{F} \rangle}. \tag{5.8}$$

When $\bar{F}[\mathbf{m}(H)]$ is greater than $\langle \bar{F} \rangle$, i.e., the average fitness of the schema is greater than the average fitness of the population, $M(H, t + 1) > M(H, t)$. Otherwise, $M(H, t + 1) < M(H, t)$. Therefore, above-average schemata increase in number, and below-average schemata decrease in number in successive generations by the selection or reproduction process. The selection process will gradually improve the fitness of the population but will not generate any new members. The processes of crossover and mutation do exactly that. Let the crossover and mutation probabilities be p_x and p_m, respectively, and l the number of bits in the chromosome. When the processes of single-point crossover and mutation are combined with the selection process, $M(H, t + 1)$ can be related to $M(H, t)$ by the following relation (Goldberg 1989):

$$M(H, t+1) \geq M(H, t) \frac{\bar{F}[\mathbf{m}(H)]}{\langle \bar{F} \rangle} \left(1 - p_x \frac{\delta(H)}{l-1} - o(H) p_m \right), \tag{5.9}$$

where $\delta(H)$ and $o(H)$ are the length and order of schema H, respectively. The preceding equation shows that for high values of mutation and crossover, the number of above-average schemata are likely to decrease, slowing down the convergence. This, however, may not necessarily be bad because it may prevent convergence to a suboptimal fitness value. This also shows that short, low-order, above-average schemata receive increasing trials in subsequent generations.

The schema theorem demonstrates how a basic GA works in a simple and elegant manner. It clearly is not based on any rigorous mathematical model and does not prove that a GA will converge to the globally optimal value of the fitness function. It does, however, demonstrate how the GA parameters affect the performance of an algorithm and therefore plays a very important role.

5.3 Problems

Practitioners using GAs have recognized both the advantages and limitations of the method. Methods of solution or improvements on the basic concept can be loosely categorized as either ad hoc and suited to the problem at hand or numerical extensions based on analogy to natural reproduction and evolution. We refer to the latter as the *pure* GA approach. While we find their extensions to the basic algorithm interesting, they are not always easy to employ or may provide only minor improvements in practice.

The primary advantage of the basic GA is that it always converges toward models with higher fitness values for reasonable crossover and mutation probabilities. Our experience in geophysical applications has shown that multipoint crossover is preferred, and using from half to twice the number of models as there are bits in the binary-coded model gives good results with a crossover probability of 0.6 and a mutation probability of 0.01. If more models are used in the population, the GA generally gives better performance, but the cost may be significantly increased.

The basic GA is quite robust. The method is not particularly sensitive to the randomly selected starting models as long as a sufficient number of models are employed. If the initial random sampling of model space is adequate, many good models will be found by the GA.

There are, however, several problems with the basic algorithm that need to be addressed. First, there is no guarantee that the optimal solution will be found. This may be disconcerting, but in many cases, good solutions may be adequate, and the basic GA with proper choice of algorithm parameters will always find many good solutions.

The algorithm proceeds toward a fit population; i.e., it converges toward the best model of the population. This convergence can be premature if a small number of models are employed. Rapid convergence to minimize computational cost may be undesirable because the model space will not be adequately searched, and the population may become homogeneous around a fitness maximum that is not near the global maximum.

Conversely, convergence toward the global maximum may be slow because some model parameters have a minor impact on the fitness. Extensive sampling of the model space often results in only minor improvements to the population's fitness at significant computational cost.

The basic algorithm is also unsatisfying in that once the models have been selected and reproduction occurs, other previously acquired information is not usually exploited.

GA purists have addressed these and other issues and have also tried to establish a mathematical framework that describes and validates this optimization approach. To address the issue of premature convergence and insensitivity of the fitness function to some model parameters, the concept of *stretching* the fitness values emerged (e.g., Goldberg 1989). During the early generations, differences in the fitness values of the population are de-emphasized. The idea is to not unduly bias the selection processes toward the good models in the population in the early generations. These good models likely correspond to local fitness maxima, and it is not desirable to have many of the other models discarded at this early point. How the stretching of the fitness values is accomplished is typically problem-dependent.

The problem of the insensitivity of the fitness to changes in the model parameters, particularly at later generations, is also problem-dependent. The optimization problem may be parameterized poorly or in such a way that some of the model parameters have little effect on the fitness as originally defined. The GA may continue its search toward a fitter population, but the convergence may be very slow. A better definition of the model parameters being searched, the extent of the search space, or the resolution may all provide solutions. Another possibility is that the definition of fitness could be changed.

One solution that has been employed successfully in GAs is to exaggerate the differences in the fitness values at the later generations so that the process of selection is biased toward the best-fit models, even though the fitness differences may be minor. This is the opposite of the de-emphasis that is desirable in the early generations. Usually a continuous generation-by-generation change in the relative shrinking and stretching of the fitness values is employed. An example of linear scaling of fitness can be found in Goldberg (1989).

An alternate scheme is tournament fitness ranking, where a discrete ranking based on the number of models in the population is employed (e.g., Baker 1987; Whitley 1989). The inclusion of desirable genetic information from previous generations is also possible in more advanced GAs (Goldberg 1989).

5.4 Combining elements of SA into a new GA

While SA has its origins in physics, GAs have their roots in biology. SA has an elegant proof of convergence to an equilibrium distribution, but this convergence may be impractical to realize in practice. GAs have no proof of convergence to the global optimum, but solutions at or very near the global optimum are usually found. Concepts from each approach are intuitively appealing, and each approach has its limitations. We now discuss ways to combine elements of each algorithm into a new hybrid algorithm.

First, we recognize that the stretching of the fitness employed in GAs plays the same role as temperature in SA. All the ideas related to temperature and temperature schedules can be readily applied to GAs. If we assume that our fitness is normalized so that unity represents perfect fitness, we can see that a temperature above 1 will cause differences in the fitness values to be de-emphasized and temperatures below 1 will cause differences in the fitness values to be exaggerated. The temperature can be changed in each generation, and the experience gained in SA on initial temperatures and rates of cooling can now all be applied to GAs.

Second, consider the evolution of models 1, 2, 3, etc. in the population as it undergoes the genetic operations of selection, crossover, and mutation. The genetic operators can be grouped together and considered a replacement for the pure random walk employed in SA. Since we are evaluating many models simultaneously in a GA, we can look between the models for guidance; i.e., we are employing either a quasi-random or pure random walk based on the probabilities of crossover and mutation. Next, we ask, Should we accept all the new models after their random walks or reject some of them? To do this, we introduce a very simple GA analogue for heredity. We simply keep track of the previous generation's models and fitness values and use these to form a temperature-dependent acceptance probability for models 1, 2, 3, etc. The result is that some of the new models formed by the GA operators will be kept, and others will be rejected. Thus we proposed (Stoffa and Sen 1991) two important modifications of a classical GA. They are as follows:

- Replace the selection probability as given in Eq. (5.4) with a temperature-dependent selection probability given by

$$p_s\left(\mathbf{m}_i\right) = \frac{\exp\left[\dfrac{F\left(\mathbf{m}_i\right)}{T}\right]}{\displaystyle\sum_{j=1}^{n} \exp\left[\dfrac{F\left(\mathbf{m}_j\right)}{T}\right]}, \tag{5.10}$$

where T is a parameter that controls the stretching of the fitness analogous to temperature used in SA.

- Use an update probability to decide whether a model from a previous generation should be used in place of a current model. Once a new model's fitness function has been evaluated, it is compared to the fitness function of a model from the previous generation selected uniformly at random and used only once. If the current model's fitness function is greater, the current model is always kept. If it is less, the previous generation's model replaces the current model with the specified update probability p_u. This update probability now plays the role of the acceptance probability as used in classical Metropolis SA.

We now re-examine the schema theorem (Eq. 5.8) with the new selection probability (Eq. 5.10) to study the effect of temperature on the selection process. From Eqs. (5.8) and (5.9), we have (Sen and Stoffa 1992a)

$$M(H,t+1)=M(H,t)\frac{\dfrac{1}{n}\sum_{i=1}^{M(H,t)}\exp\left[\dfrac{F(\mathbf{m}_i)}{T}\right]}{\dfrac{1}{M(H,t)}\sum_{k=1}^{n}\exp\left[\dfrac{F(\mathbf{m}_k)}{T}\right]}. \tag{5.11}$$

When the temperature T in the preceding equation becomes sufficiently small, we have

$$M(H,t+1)\cong n. \tag{5.12}$$

Thus for low values of T, schemata with above-average fitness (which may not necessarily be the optimal fitness) will be extensively reproduced. Also,

$$\lim_{T\to\infty}\exp\left[\frac{F(\mathbf{m}_i)}{T}\right]=1, \tag{5.13}$$

such that

$$M(H,t+1)=M(H,t). \tag{5.14}$$

Thus a high value of T retains diversity among the members of the population, allowing for improvement in subsequent generations.

Figure 5.7 summarizes the new algorithm. It also indicates that the algorithm can be terminated before the specified maximum number of generations. This can occur when a fitness greater than some specified fitness is achieved or when the population becomes nearly homogeneous, as measured by the ratio of the mean fitness of the population to the best fitness of the current generation.

The resulting algorithm can be considered a modified version of the basic GA, with the ad hoc fitness stretching replaced by a temperature-dependent probability and with the inclusion of some heredity information based on a model's fitness compared with its fitness in the preceding generation.

5.5 A mathematical model of a GA

Attempts have been reported (e.g., Goldberg and Segrest 1987; Eiben *et al.* 1991; Davis and Principe 1991; Mahfoud 1991; Horn 1993; etc.) to model GAs by means

---

Genetic algorithm

1) For generation 1 (g=1), select an ensemble of N models at random,

   $\mathbf{m}_{g,j}, j$=1, . . , N ; g=1,..., NG.

2) Evaluate the objective (fitness) function $f_{g,j}$ for each model:

   if $g = NG \rightarrow$ STOP

   $$\text{let } F_g = \max[f_{g,j}] \; j\text{=}1, N \text{ and } \overline{F} = \sum_{j=1}^{N} f_{g,j}/N \;\;,$$

   if $F_g >$ some limit $\rightarrow$ STOP

   if $\overline{F}/F >$ some limit $\rightarrow$ STOP

3) For g>1 (all but the first generation) compare the previous generation's objective functions,

   $f_{g-1,j}$ with the current generation's objective functions, $f_{g,j}$

   if $f_{g,j} > f_{g-1,j}$ accept $\mathbf{m}_{g,j}$

   if $f_{g-1,j} > f_{g,j}$ then evaluate $p_{up} = \exp((f_{g,j}-f_{g-1,j})/T_A)$, or fxed by the algorithm

   if $p_{up} >$ rand [0, 1]

   $\mathbf{m}_{g,j} = \mathbf{m}_{g-1,j}$

   $f_{g,j} = f_{g-1,j}$

   else

   accept $\mathbf{m}_{g,j}$ and $f_{g,j}$

   end if

4) Select pairs of models for reproduction in proportion to their fitness, $f_{g,j}$,

   $$p_s^j = \frac{\exp\left(f_{s,j}/T_s\right)}{\displaystyle\sum_{k=1}^{N} \exp\left(f_{s,k}/T_s\right)} \quad.$$

5) For each pair apply the genetic operators:

   crossover with probability p_x, e.g., 0.6

   mutation with probability p_m, e.g., 0.1

6) Lower the temperature, go to step (2) and repeat the procedure.

Figure 5.7 A step-by-step algorithm for genetic inversion.

of Markov chains. Most of these are either restrictive or lack mathematical rigor. Vose and Liepins (1991) first proposed a meaningful mathematical model of a simple GA that was analyzed using finite Markov chains (Nix and Vose 1992). The Vose and Liepins (1991) model recognizes the collection of models (population) in a generation as a *state* that undergoes the processes of selection, crossover, and mutation to generate a new population of models or a new state. Thus the genetic processes of selection, crossover, and mutation need to be explicitly taken into account in the definition of a transition probability matrix that characterizes the transition from one state to another. In the following we will closely follow the derivations in Vose and Liepins (1991) and Nix and Vose (1992) to obtain an expression for the transition probability matrix that takes into account all three genetic processes.

Let

l = the length of a chromosome or number of bits in a model;
$r = 2^l$, the number of possible models (solutions);
n = population size; and
N = total number of possible populations.

It can be shown (Nix and Vose 1992) that

$$N = {}^{n+r-1}C_{r-1}. \tag{5.15}$$

That is, N is the total number of combinations C of $(n + r - 1)$ objects taken $(r - 1)$ objects at a time. At this stage we also define an $r \times N$ matrix $\mathbf{Z}$ such that each column of the matrix represents a population of size n. These columns will be associated with the population at each generation.

Assume, for example, that we have a problem in which each model is represented by only 2 bits, i.e., $l = 2$, and we are considering only one model ($n = 1$) in the population. For a model with only 2 bits, the total number of possible models $r = 2^2 = 4$, and they are

$$\begin{aligned}
\mathbf{m}_0 &= \{0 \ 0\}, \\
\mathbf{m}_1 &= \{0 \ 1\}, \\
\mathbf{m}_2 &= \{1 \ 0\}, \\
\mathbf{m}_3 &= \{1 \ 1\},
\end{aligned} \tag{5.16}$$

Note that in our notation for models, the number in the subscript denotes the decimal value of the bit string. Thus $\mathbf{m}_0$ indicates that all the bits are 0, whereas $\mathbf{m}_{r-1}$ indicates that all the bits are 1.

Also, for this problem, $N = 4$, and the matrix $\mathbf{Z}$ is a 4×4 matrix given by

$$\mathbf{Z} = \begin{bmatrix} z_0^0 & z_1^0 & z_2^0 & z_3^0 \\ z_0^1 & z_1^1 & z_2^1 & z_3^1 \\ z_0^2 & z_1^2 & z_2^2 & z_3^2 \\ z_0^3 & z_1^3 & z_2^3 & z_3^3 \end{bmatrix}. \tag{5.17}$$

Each column of $\mathbf{Z}$ denoted by the vector $\mathbf{z}$ characterizes a population; i.e.,

$$\mathbf{z}_j = \begin{bmatrix} z_j^0, z_j^1, z_j^2, z_j^3 \end{bmatrix}^T, \tag{5.18}$$

etc., and each element Z_j^i where i is the model index and j is the population index, is the number of occurrences of model $\mathbf{m}_i$ in population j. Thus Z_0^2 is the number of occurrences of model $\mathbf{m}_2$ in population 0. Therefore, for the 2-bit, one-model GA example, $\mathbf{Z}$ becomes

$$\mathbf{Z} = \begin{bmatrix} 1 & 0 & 0 & 0 \\ 0 & 1 & 0 & 0 \\ 0 & 0 & 1 & 0 \\ 0 & 0 & 0 & 1 \end{bmatrix}. \tag{5.19}$$

Since we restricted ourselves to one model in each generation, each entry in the matrix is either 0 or 1 so as to define four unique populations. The sum of the elements in each column of $\mathbf{Z}$ is equal to n (in this case 1), the population size.

If we assume that a population is a *state* characterized by a column of the matrix $\mathbf{Z}$, we need to define or determine the form of a transition probability matrix $\mathbf{P}$ with entries P_{ij} (an $N \times N$ matrix), which is the probability of being in the population or state j at generation $k + 1$ given that we are in state i at generation k. Clearly, the outcome at generation $k + 1$ depends only on the preceding generation, and therefore, the process can be modeled with a Markov chain.

Recall that in a SA the transition matrix is given by a generation and an acceptance probability, where the generation probability is often a uniform distribution. It is easy to follow the concept because each model is considered to be a state, and a transition implies transition from one model to another.

The GA process involves the entire population in a generation, and the generation of a new state depends on three processes, namely, selection, crossover, and mutation, and all three processes will have to be taken into account in defining a transition probability. Recall that the vectors $\mathbf{z}_i$ and $\mathbf{z}_j$ represent populations in generations i and j, respectively, and Z_j^k gives the number of model occurrences of $\mathbf{m}_k$ in generation j.

Clearly, from a population of size n, the number of ways of choosing Z_j^0 models of $\mathbf{m}_0$ is given by the combination

$$^nC_{z_j^0}.$$

Now, for $\mathbf{m}_1$, the number of ways of choosing Z_j^1 numbers of $\mathbf{m}_1$ is given by

$$^{n-z_j^0}C_{z_j^1},$$

and so on.

Thus the total combination of all the models is given by the product of the combination terms as

$$
{}^{n}C_{z_j^0} \cdot {}^{n-z_j^0}C_{z_j^1} \cdots {}^{n-z_j^0-z_j^1-\cdots-z_j^{r-2}}C_{z_j^{r-1}} = \frac{n!}{z_j^0!\left(n-z_j^0\right)!} \frac{\left(n-z_j^0\right)!}{z_j^1!\left(n-z_j^0-z_j^1\right)!} \cdots \frac{\left(n-z_j^0-z_j^1-\cdots-z_j^{r-2}\right)!}{z_j^{r-1}!\left(n-z_j^0-\cdots-z_j^{r-2}-z_j^{r1}\right)!}
$$

$$
= \frac{n!}{z_j^0!\,z_j^1!\cdots z_j^{r-1}!}.
$$

(5.20)

Now let $p_i(k)$ represent the probability of producing model $\mathbf{m}_k$ in the next generation, given the current generation i. In our notation Z_j^k is the number of occurrences of model $\mathbf{m}_k$ in generation j. Thus the probability of Z_j^j occurrences of $\mathbf{m}_k$ is

$$
p_i(k)^{z_j^k}.
$$

Thus the transition matrix P_{ij}, which gives the probability of transition from generation i to generation j, is given by

$$
P_{ij} = \frac{n!}{z_j^0!\ z_j^1!\cdots z_j^{r-1}!} \prod_{k=0}^{r-1}\left\{p_i(k)\right\}^{z_j^k}
$$

$$
= n!\prod_{k=0}^{r-1}\frac{\left\{p_i(k)\right\}^{z_j^k}}{z_j^k!}.
$$

(5.21)

This description of the transition probability is far from being complete because it depends on the probability $p_i(k)$, which should include the effects of all three processes, namely, selection, crossover, and mutation. Vose and Liepins (1991) gave a detailed derivation of $p_i(k)$ that is also outlined in Nix and Vose (1992). We refer the reader to these two papers for details and reproduce here only the final result.

In order to find the probability of occurrence of a model (say $\mathbf{m}_k$) in a generation, we need to consider the following two points:

- In any generation i, the kth component of the vector $\mathbf{z}_i$ denoted simply as Z_i^k gives the number of occurrences of model $\mathbf{m}_k$ in generation i. Let F be an $l \times l$ diagonal matrix, each diagonal element of which gives a non-negative fitness of a model; i.e.,

$$
\mathbf{F} = \begin{bmatrix} F(\mathbf{m}_0) & & & & \\ & F(\mathbf{m}_1) & & & \\ & & \cdot & & \\ & & & \cdot & \\ & & & & \cdot & \\ & & & & & F(\mathbf{m}_{l-1}) \end{bmatrix}. \tag{5.22}
$$

Therefore, the quantity

$$
\left(\frac{\mathbf{F}\mathbf{z}_i}{|\mathbf{F}\mathbf{z}_i|} \right)_k \tag{5.23}
$$

gives the probability of selection of a particular model $\mathbf{m}_k$. That is,

$$
p_s(\mathbf{m}_k) = \frac{F_{kk} z_i^k}{\displaystyle\sum_{j=0}^{l-1} f_{jj} z_i^j}. \tag{5.24}
$$

Thus Eq. (5.23) essentially describes the effect of the selection process on the population given by the vector $\mathbf{z}_i$.

- Once a model, say $\mathbf{m}_k$, has been selected for recombination, the next step would be to find the probability of occurrence of the model after recombination. Vose and Liepins (1991) described crossover and mutation by means of an operator M. This involved a very careful analysis of crossover and mutation by means of bit-wise exclusive-or and logical-and operations and the definition of a permutation function in terms of these logical operators. Please see Vose and Liepins (1991) for a detailed derivation of the permutation function and the operator M.

Thus the derivation of $p_i(k)$ is complete. The outcome of the selection, as given by Eq. (5.23), should now be acted on by the operator M to give $p_i(k)$; i.e.,

$$
p_i(k) = M\left(\frac{\mathbf{F}\mathbf{z}_i}{|\mathbf{F}\mathbf{z}_i|} \right)_k. \tag{5.25}
$$

The transition probability is now given as

$$
P_{ij} = n! \prod_{k=0}^{r-1} \frac{\left\{ M\left(\frac{\mathbf{F}\mathbf{z}_i}{|\mathbf{F}\mathbf{z}_i|} \right)_k \right\}^{z_j^k}}{z_j^k!}. \tag{5.26}
$$

Now that a transition probability matrix that characterizes transition from a population i to a population j can be written down following the approach of Nix and Vose (1992), we ask the following questions:

- How are the populations distributed if the genetic processes of selection, cross-over, and mutation are repeated a large number of times; i.e., what is the asymptotic steady-state or equilibrium distribution?
- Does a GA converge to a particular population after a large number of transitions, and can the convergence behavior of a GA be predicted?

To determine the steady-state distribution, we need to evaluate Eq. (5.26) in the limit that the number of transition is very large (infinite). Such an equilibrium distribution is independent of the initial population because, for a non-zero mutation rate, the Markov chain may be considered ergodic. However, determination of such an equilibrium distribution in its explicit form (similar to SA) as a function of fitness and algorithm parameters, such as population size, and mutation and cross-over rates remains an unsolved (and may be impossible) task.

Nix and Vose (1992) and Vose (1993) further analyzed Eq. (5.26) in the limit that the population size goes to infinity and concluded that the convergence behavior of a real GA can be accurately predicted if the finite population is sufficiently large. When the population size increases, then the GA asymptotically reaches a population that corresponds to a unique (attracting) fixed point.

At this stage we will re-examine the Stoffa and Sen (1991) algorithm. The selection probability, as given in Eq. (5.10), is easy to incorporate in the model of Vose and Liepins (1991) simply by substituting Eq. (5.10) into Eq. (5.22). However, most GAs lack the acceptance criterion, such as the update probability introduced by Stoffa and Sen (1991). Recall that in the update process, after a new model's fitness function has been evaluated, it is compared with the fitness function of the same model of the population from the preceding generation. This process is repeated for each member (model) of the population. Since the Vose and Liepins (1991) model considers the entire set of models as a state in a Markov chain and the *update* process is applied to each model in the generation, the inclusion of update in the Vose and Liepins (1991) model is not a trivial or a straightforward task. We will therefore consider a simple, albeit less rigorous, model of GA with the aim of comparing the GA of Stoffa and Sen (1991) with SA.

To describe the algorithm, we will follow a particular member of the population through different generations. In this algorithm, a trial consists of the processes of selection, crossover, mutation, and update. Thus this is a two-step procedure. First, a model is generated by the processes of selection, crossover, and mutation. Then its fitness is compared with that of a model from the preceding generation.

The new model is always accepted if the fitness of the new model is higher. When the fitness is lower, it is accepted with an update probability given by the well-known Metropolis acceptance criterion as described in Chapter 4 in the context of SA. Following the procedure as described in the context of Metropolis SA (see Chapter 4), we can express the probability of transition from a model i to a model j as a product of two probabilities; i.e.,

$$P_{ij} = G_{ij} A_{ij}, \qquad (5.27)$$

where G_{ij} is the generation probability and A_{ij} is the acceptance probability. These are given by the following equations:

$$G_{ij} = G_j = f_1(p_x) f_2(p_m) p_s^j = f_1(p_x) f_2(p_m) \frac{\exp\left[\dfrac{F(\mathbf{m}_j)}{T_s}\right]}{\displaystyle\sum_{k=1}^{n} \exp\left[\dfrac{F(\mathbf{m}_k)}{T_s}\right]}, \qquad (5.28)$$

and

$$A_{ij} = \exp\left\{\frac{\left[F(\mathbf{m}_i) - F(\mathbf{m}_j)\right]^+}{T_A}\right\}, \qquad (5.29)$$

where f_1 and f_2 are functions of the crossover and mutation probabilities, respectively. Depending on their values, we may consider them to be generators of random models. Note the existence of the control parameter temperature in both the selection and acceptance probabilities T_S and T_A and that they can take different values.

The model described by these equations is very similar to that used in describing Metropolis SA except that the generation probability in Metropolis SA is uniform, whereas that used in the GA is biased through the selection criterion given by Eq. (5.10) and the process of crossover ($p_x \neq 0$). Thus the GA selection process, like that of heat bath SA and SA without rejected moves (SAWR), is more global than a pure Metropolis SA. It is intuitively obvious that such an approach is desirable to avoid the high rejection-to-acceptance ratio in Metropolis SA. This, however, causes problems when we attempt to describe the GA in terms of a Markov chain. Although for a finite mutation rate the Markov chain may be considered ergodic (i.e., there is finite probability of reaching every state from every other state), the selection probability as given by Eq. (5.10) and used in Eq. (5.28) is controlled by the partition function in the denominator, which takes on different values for

different generations. This can be sidestepped by saying that for a given mutation rate and a large population size, the sum in the denominator remains nearly a
constant at different generations. Under this more restrictive condition, an analysis
similar to that used in Metropolis SA or SAWR can be carried out, and an expression for the steady-state distribution (which again turns out to be a Boltzmann distribution) can be achieved.

However, consider another situation when the selection temperature T_S is very
high. A model is then selected uniformly at random from the population. In such
a situation, if the crossover rate is zero and mutation is high, the Stoffa and Sen
(1991) GA is the same as running several independent Metropolis SAs in parallel. This, however, does not serve our goal of improving computational speed.
However, if we consider a high T_S, moderate crossover ($p_x = 0.5$), and moderate
mutation ($p_m = 0.5$), then the crossover mixes models between different parallel
SAs, and mutation allows for a local neighborhood search. We will show numerically that such an approach does indeed converge to a Gibbs distribution at a constant acceptance temperature. We call this new approach a *parallel Gibbs sampler*
(PGS) and show numerical results for geophysical applications in Chapter 8.

5.6 Multimodal fitness functions, genetic drift, GA with sharing, and repeat (parallel) GA

The classical GA based on three operators acting on simple haploid chromosomes
has been used successfully in many optimization problems. Their performance
can be further improved by properly stretching and contracting the fitness function
(Stoffa and Sen 1991; Sen and Stoffa 1992a) using concepts from SA. However,
there remains an inherent problem of poor performance of the algorithm on multimodal (multiple-peaked) fitness functions where the peaks are nearly equal in
height. Goldberg and Richardson (1987) showed that when a GA is used on a
multimodal function, it converges to one peak or the other. That is, if we have a
fitness function with many peaks of nearly equal height (this can also occur due to
numerical implementation of a fitness function that has a high constant value for a
range of models) at different points in the model space, it converges to one point
in the model space or the other. Therefore, if the GA is allowed to converge and
we look at the models picked at the last generation, we find that almost 95 percent
of the models are identical. This is caused by the process of crossover, whereas
the mutation is only able to maintain a small amount of diversity in the final population. Such a phenomenon will cause problems in geophysical inversion, where
the aim is not only to derive a best-fit answer but also to characterize the uncertainty in our answer. Much of the discussion in Chapter 2 is devoted to this topic.

Particularly when the problem is underdetermined, several non-unique solutions are possible, and we would like to know about them.

In our geophysical applications, we do not know the shape of the fitness function unless we evaluate the fitness function at each point in the model space – something we always want to avoid. Usually we hope that the function has several small peaks and a large peak (the global maximum). Although a GA is not guaranteed (asymptotically or otherwise) to converge to the global maximum, our experience with such a problem has been that with proper choice of the GA parameters and by incorporating SA concepts, it is possible in most cases to arrive very close to the global maximum of the fitness function. However, this alone may not be enough in many situations.

The fundamental theorem of GA, which shows that an exponentially increasing number of trials will be given to the observed best schemata, assumes that the population size is infinite. In a finite population, even when there is no selective advantage of one peak over the other, the population converges to one population or the other in a finite time (Goldberg and Segrest 1987). This problem has been called *genetic drift* by the geneticists. This problem is rather perplexing, especially when compared with natural examples. For example, when several species have nearly the same fitness value, nature does not converge to a single species. In nature, different species do not go head-to-head. Instead, they exploit separate niches (environmental factors) in which other organisms have little or no interest. A well-developed literature on niche and speciation exists in biology. In artificial GAs, this can be achieved by a method called *sharing*. Sharing can be done in two ways: phenotypic sharing (according to decoded parameter values) and genotypic sharing (according to hamming distance between chromosomes). Goldberg and Richardson (1987) successfully used these concepts in GAs in their application to a multimodal fitness function in one dimension. Their method involves using a *sharing function* obtained by measuring the similarity between different models in a generation. The sharing function derates fitness according to the number and closeness of neighboring points. Although *GA with sharing* was applied successfully to a single-dimensional multimodal fitness function, we found that it takes many generations to attain a high value. For example, the convergence was very slow in our application to inversion of plane-wave seismograms (Sen and Stoffa 1992a). Multiparameter sharing methods have been discussed by Deb and Goldberg (1989).

One other method that can be used to circumvent the problem of genetic drift is a parallel GA (Cohoon *et al.* 1987) based on the concept of *punctuated equilibria* (PE) (Eldredge and Gould 1972). Punctuated equilibria is a theory used to resolve certain paleontological dilemmas in the geologic record. PE is based on two principles: allotropic speciation and stasis. Allotropic speciation involves rapid evolution of new species after being geographically separated. This involves a small subpopulation of a species becoming segregated into a new environment. Stasis implies that after equilibrium is reached in an environment, there is very little drift away from the genetic composition of the species. Although the concept of PE in paleontology is highly debated, it has been used successfully in developing parallel GAs (Cohoon *et al.* 1987). The parallel GA assigns a set of *MM* models to each of *NP* processors, for a total population of (*MM · NP*). Thus each processor corresponds to a disjoint environment. This is also similar to making *NP* independent repeat GA runs, each with a different starting population of size *MM*. In either case, we expect to see the emergence of some very fit species. Then a catastrophe occurs, and environments change. This is simulated by mixing models between different environments.

Similar concepts were used in seismic waveform inversion (Stoffa and Sen 1991; Sen and Stoffa 1992a). Several repeat GAs were run with a fixed number of different randomly generated models. The results from all these runs were collected, and the distributions of the models was examined to characterize uncertainty.

Making parallel or repeat runs has several advantages. First, if the fitness function is multimodal with several peaks of nearly the same height at different points (not necessarily closely spaced), we can sample many models close to each peak in each of the runs. This is useful even when the fitness function has one highest peak with several smaller peaks. Although a GA with proper choice of GA parameters can be made to reach very close to the global maximum of the fitness function, each run will sample a slightly different portion of the model space, resulting in several closely spaced points around the global maximum. Also, it is more likely by randomly selecting the initial models in repeat runs that the sampling will be a better representation of the entire population. Thus the chances of getting trapped into a local maximum are greatly reduced, even when a small population size is used.

5.7 Uncertainty estimates

As mentioned in earlier chapters, the objective of geophysical inversion is not only to find a best-fit model but also to characterize the uncertainty in the result. This is usually characterized by the marginal posterior probability density function and different measures of dispersion such as the mean, covariance, etc. (Eqs. 2.147 through 2.150).

An unbiased estimate of these integrals can only be obtained by importance sampling, in which the models are drawn from the PPD $\sigma(\mathbf{m} \mid \mathbf{d}_{obs})$. Since SA is statistically guaranteed to converge to the Gibbs distribution after a large number

select a population of models $\mathbf{m}_i$, i = 1, ..., n with fitness $E(\mathbf{m}_i)$
loop over generation
- loop over models (i = 1, n)
- • $\mathbf{m}_{i+n} = \mathbf{m}_i + \Delta\mathbf{m}$
- $\Delta\mathbf{m}$ is drawn from a Gaussian distribution around $\mathbf{m}_i$ with $E(\mathbf{m}_i)$ scaled as a variance
- • evaluate fitness $E(\mathbf{m}_{i+n})$
- end loop

  now we have $2n$ models

- loop over models (i = 1, 2n)
- • loop over models (j = 1, k; k<n)
- • • k1 = U [1, 2n]
- • • u1 = U [0, 1]
- • • $ff = E(\mathbf{m}_{k1})/[E(\mathbf{m}_{k1}) + E(\mathbf{m}_i)]$
- • • if (u1.gt.ff) then
- • • • wt = 1.0
- • • else
- • • • wt = 0.0
- • • end if
- • • w(i) = w(i) + wt
- • end loop
- rank $\mathbf{m}_i$ in the decreasing of w(i)
end loop

Figure 5.8 A pseudo-Fortran code for EP. (Based on Fogel, 1991.)

of trials, evaluation of Eq. (2.150) using SA has often been preferred. On the other hand, GAs have often been criticized for the lack of a rigorous mathematical description. Since the models sampled by GA are not distributed according to $\sigma(\mathbf{m} \mid \mathbf{d}_{obs})$ (except possibly for more restrictive algorithms, as mentioned in Section 5.5), it is expected that estimates of such integrals will be biased. We will show in Chapter 8 how a parallel Gibbs sampler and multiple GAs can be used to obtain approximate values of these integrals.

5.8 Evolutionary programming – a variant of GA

So far our discussions of GAs were largely based on the algorithms introduced by Holland (1975) and some related forms. However, the idea of simulating natural environments on computers dates back to the work of Friedberg (1958), who attempted to gradually improve a computer program. Fogel *et al.* (1966) used evolutionary programming (EP) in system identification in which finite-state machines were considered to be organisms. Of the organisms that solved for a target function, the best were allowed to reproduce. The parents were mutated to create offspring. However, Holland (1975) proposed a GA that used many new concepts, such as coding, crossover, etc. Later there was renewed interest in EP, as evidenced by the work of Fogel (1991). Fogel (1991) pointed out that the crossover operation, which is unique to GAs, leads to one of the fundamental difficulties of GAs. That is, it may cause premature convergence. This is so because after successive generations, the entire population converges to a set of codings such that the crossover no longer generates any new chromosomes. This may happen even before finding an optimal solution. Although mutation allows for diversity, the mutation rate is usually low, so practically no improvement can be attained in the final generations of the algorithm.

Fogel and Fogel (1986) re-examined the EP algorithm, and Fogel (1988) applied it to the traveling salesman problem. Fogel (1991) also outlined an EP appropriate for function optimization or geophysical inversion. An outline of the algorithm based on Fogel (1991) is given in Figure 5.8.

Like a GA, an EP starts with a population of models (let n = the number of models in the population), and the fitness function is evaluated for each of the models in the population. Next, an equal number of offspring are generated by perturbing each member of the population by Gaussian mutation. This means that each model is given a perturbation that is drawn from a Gaussian distribution whose variance is obtained by properly scaling the fitness function of the model. Thus a poor model is allowed to go far from its current position, whereas improvements are sought in the close neighborhood of a good model. Next, tournament selection, as described in Figure 5.7, is applied, and a score is assigned to each of the $2n$ models. This is done as follows: Each model is allowed to compete with a subset (e.g., k) of the $2n$ models drawn uniformly at random based on their fitness. If the current model wins, it scores 1; otherwise, it is assigned a score of 0. The competition is probabilistic, and thus worse models have some finite probability to win. Note, however, that no temperature-type control parameter is used in the algorithm. The maximum score that any model can obtain is k. Next, the $2n$ models are ranked in decreasing order of their fitness. The first n of the $2n$ models are now considered parents, Gaussian:mutations are allowed to create new n offspring, and the process is repeated until the population converges to a high fitness value.

Fogel (1991) gave several examples of function optimization with EP and showed that in many cases EP performed better than GAs. Again, it is very difficult to say whether such performance can be expected for every optimization problem. Since this method does not use the crossover operation, it may result in a more diverse population than a single GA run, which may be quite useful for geophysical inversion.

5.9 Summary

GAs are a novel and interesting approach to optimization problems. Unlike SA, the commonly used GA still lacks a rigorous mathematical basis. Its innovative appeal lies in its analogy with natural biologic systems and the fact that even a simple GA can be applied to find many good solutions quickly. We have shown how some aspects of GA, e.g., stretching the objective function, can be replaced by incorporating the concept of temperature into the probabilities for the selection process. Also, the genetic concept of heredity can be replaced with the Metropolis criterion to decide whether each new model should be replaced by its predecessor. The modification to a classical GA allows the algorithm to be viewed as either an analogue to biologic reproduction or as an approximation to parallel Metropolis SA. In either case, the principal distinction is the introduction of the crossover operator, which allows the models in the distribution to share information. This is an advantage because the information common to good models propagates to the next generation more efficiently and a disadvantage because premature convergence can occur. Each model of a given generation can be evaluated independently, and each run can be made independently. This makes GAs ideally suited for implementation in parallel computer architectures.

6

Other stochastic optimization methods

In Chapters 4 and 5 we described in great detail two of the most popular global optimization methods in geophysics. Simulated annealing (SA) uses a random walk and an update rule to decide on the next move. On the other hand, genetic algorithms (GAs) use a population of models and rules of selection, mutation, and update to generate the next set of members of the population. In this chapter we will review some other population-based global optimization methods that are becoming increasingly popular in geophysical applications. These include the neighborhood algorithm (NA), particle swarm optimization (PSO), and simultaneous perturbation stochastic approximation (SPSA).

6.1 The neighborhood algorithm (NA)

NA was introduced by Sambridge in 1999 with an example of the application of geophysical inversion. Recall that both SA and GAs make several random moves that are rejected. The primary motivation behind the development of this algorithm was to address this issue. Sambridge (1999a) asked the fundamental question, How can a search for new models be best guided by all previous models for which the forward problem has been solved (and hence the data misfit value evaluated)? To address this, NA first divides the model space into a set of regions, in each of which the objective function value is assumed to be a constant. All the points within a region are assigned a constant value for the objective function – a value that was computed for a point closest to it. There can be many different ways to measure the closeness and to define these regions. NA uses a method called *Voronoi tessellation* for this purpose, which has been used in many different disciplines.

6.1.1 Voronoi diagrams

Voronoi diagrams or Voronoi tessellation is a very simple, intuitive concept that has been used in a wide variety of applications. Simply stated, a *tessellation* or *tiling* is a process by which a shape is created in a 2D plane or a 3D surface that is repeated a large number of times. In our application, such a tile will contain all points in the model space of identical values of an objective function. Voronoi diagrams or Voronoi tessellation is a special type of decomposition of a metric space determined by a specific discrete set of points called *Voronoi centers* (Okabe *et al.* 1999). The original idea of this type of tessellation was introduced by Descartes in his treatment of cosmic fragmentation by stars, which is perhaps not widely known. Since he did not describe the rules in detail for generating these diagrams, he is not credited with their invention. Descartes was followed by Dirichlet and Voronoi (Okabe *et al.* 1999, p. 6), who considered a special form of Voronoi diagrams. The first application of Voronoi diagrams was in crystallography; since then, they have been applied in many other fields, including astronomy, archaeology, meteorology, physics, physiology, and urban planning.

The simplest approach to describing a Voronoi diagram is to consider a practical problem of placement of several supermarkets (e.g., a chain) in a city in the most profitable manner. One criterion that every customer will consider is the proximity of the supermarket to his or her residence. Thus people from all the residences closest to a particular supermarket will most likely visit that supermarket. In other words, a region that maps all the residences closest to a supermarket is the Voronoi cell, and the center of the Voronoi cell is the location of the supermarket.

Formally, a Voronoi diagram can be defined as follows (e.g., Okabe *et al.* 1999): Given a set S of n distinct points in R_d, a Voronoi diagram is the partition of R_d into N distinct polyhedral regions $V(\mathbf{p})$. Each region $V(\mathbf{p})$, called the *Voronoi cell* of point $\mathbf{p}$, is a set of points in R_d that are closer to $\mathbf{p}$ than any other arbitrary point $\mathbf{q}$ in the set S or, more precisely,

$$V(\mathbf{p}) = \{\mathbf{x} \in R^d ; dist(\mathbf{x}, \mathbf{p}) \leq dist(\mathbf{x}, \mathbf{q}) \, \forall \mathbf{q} \in S - \mathbf{p}\}, \tag{6.1}$$

where *dist* is the Euclidean distance function.

The dual graph of Voronoi tessellation is called the *Delaunay triangulation*. There exist several methods for constructing Voronoi tessellation; some of them are described in the text by Okabe *et al.* (1999). Figure 6.1 shows a Voronoi tessellation generated from Voronoi centers using two measures of distance, namely, L_1 and L_2 (Dimri, Srivastava, and Vedanti 2012).

6.1.2 Voronoi diagrams in SA and GA

The basic ideas behind using Voronoi diagrams in SA, GAs, and subsequently, NA are described in Sambridge (1999a). Let us assume that we have a set of n randomly chosen models $\mathbf{m}$ with their corresponding objective functions $E(\mathbf{m})$. We divide the entire model space into $\mathbf{n}$ regions, each of which is assigned a constant

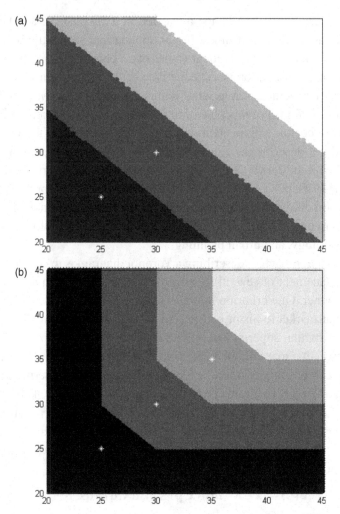

Figure 6.1 Tessellation using (a) L_2 and (b) L_1 distance functions. White stars are Voronoi centers used to generate the model. (From Dimri and Srivastava 2005.)

objective function value. In other words, we generate a Voronoi diagram; the key parameter to generating this is to measure the distance between a test point in the model space and that of a Voronoi cell (one of the randomly drawn models). We can use our standard L_2 norm to measure the distance between two points $\mathbf{m}_a$ and $\mathbf{m}_b$ given by

$$\left\| \mathbf{m}_b - \mathbf{m}_a \right\| = \left(\mathbf{m}_b - \mathbf{m}_a \right)^T C_M^{-1} \left(\mathbf{m}_b - \mathbf{m}_a \right). \tag{6.2}$$

All the parameters used in this equation were defined in Chapter 2. Note that any other norm suitable for a particular application can be used to measure the

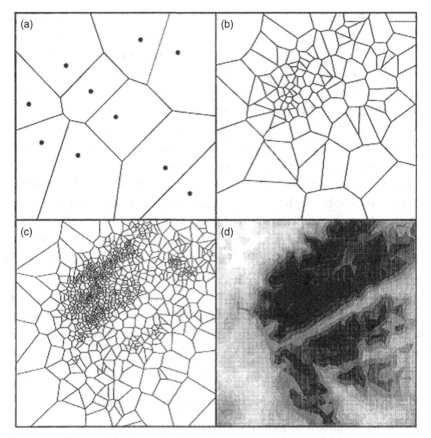

Figure 6.2 Voronoi diagrams for different numbers of Voronoi centers (a: 10 cells; b: 100 cells, and c: 1,000 cells). (d) Contour plot of the objective function. (From Sambridge 1999a.) Note that by preferential choice of Voronoi center distribution, it is possible to generate a fairly good approximation of the objective function in multidimensional space.

distances. Figure 6.2 (from Sambridge 1999a) displays Voronoi diagrams for an increasing number of Voronoi centers (10, 100, and 1,000). A contour plot of the objective function is also shown in Figure 6.2d. Note that with selective sampling (more points in the regions of changing landscape of the objective function), one can generate fairly good representation of the objective function in multidimensional space.

Let us consider a GA, for example (see Chapter 5 for details). Note that at each generation of a GA, we create a new population of models (say *NP* number of models) using genetic operators, namely, selection, crossover, and mutation. Now we need to evaluate cost function values for all these models. Note that within each generation, we generate several models, some of which are exactly the same as those existing in the preceding generation, whereas others are only slightly different. Thus, depending on the distance between the models, one may choose to simply use the cost function corresponding to the Voronoi cell and completely avoid the forward model evaluation. In principle, this approach provides a way to avoid repeat forward model calculations. In practice, the nature of the forward problem would dictate how effective this procedure can be.

Recall that in a Metropolis SA, a random model is first drawn from a proposal distribution, and its objective function is evaluated. The second step involves perturbing the current model and accepting or rejecting it based on the Metropolis criterion (see Chapter 5). The procedure of model generation and acceptance is repeated a large number of times. Once again, the idea of Voronoi cells can be used here. Before a new model's function evaluation is carried out, its closeness to an existing Voronoi cell can be examined, and a decision can be made whether new forward modeling must be done. Similar ideas can also be implemented in a heat bath–based SA.

Although Sambridge (1999a) outlined the basic framework of using these ideas, no application of Voronoi-based SA and GAs has been reported. Note that many applications do use these concepts in SA and GAs without noting the explicit use of Voronoi cells.

6.1.3 Neighborhood sampling algorithm

The NA can be summarized as follows (Sambridge 1999a):

1. Generate an initial set of n_s models uniformly in parameter space.
2. Calculate the misfit function for the most recently generated set of n_s models, and determine the n_r models with the lowest misfit of all models generated so far.
3. Generate n_s new models by performing a uniform random walk in the Voronoi cell of each of the n_r chosen models (i.e., n_s/n_r samples in each cell).
4. Go to step 2.

First, n_s numbers of models (Voronoi centers) are drawn at random from a predefined model space, and objective function values for these models are computed. Next, Voronoi cells are constructed using a measure of distance defined in Eq. (6.2), and every point in the Voronoi cell is assigned an objective function value that is the same as that at the Voronoi center. During this process, we essentially generate equi-objective function surfaces. Of the n_s models drawn so far, n_r models with the lowest misfit are chosen. Remaining $(n_s - n_r)$ models are rejected. Now, within each of these n_r Voronoi cells, n_s/n_r new models are drawn, resulting in a new set of n_s models. Once again, n_r models with the lowest misfit are chosen, and remaining $(n_s - n_r)$ models are rejected. This process is then repeated until a predefined convergence criterion is attained. Following are the advantages of NA:

- It uses previous samples to determine the next step of the search, or in other words, the algorithm is self-adaptive.
- The algorithm requires only two tuning parameters, namely, n_s and n_r.
- The process generates an ensemble of models with low misfit values; thus the method can be used for uncertainty quantification as well (Sambridge, 1999b).

Sambridge (1999a, 1999b) and many other subsequent publications provide examples of applications of NA to geophysical inversion problems, including receiver function inversion, earthquake location, etc.

6.2 Particle swarm optimization (PSO)

The PSO method is a stochastic evolutionary computation technique (Kennedy and Eberhart 1995) used in optimization that was inspired by observation of social behavior of individuals (called *particles*) in nature, such as bird flocking and fish schooling. PSO was viewed as a combination of two processes: a long-term process, the biologic evolution that has spanned thousands of years, and a short-term process, the neural processing that occurs on the sub-millisecond time scale. The PSO algorithm uses ideas that are very similar to those used in evolutionary computation techniques such as GAs, as described in Chapter 5. As with GAs, a PSO also starts with a population of random solutions and searches for optima by updating generations. However, unlike GAs, a PSO does not make use of evolution operators such as crossover and mutation. In a PSO, the potential models, called *particles*, navigate the model space by following the current optimal particles. The algorithm draws inspiration from social psychology in that each solution uses its own experience together with the experiences of its neighbor to generate a new trail solution. The algorithm has been successfully used in many different science and engineering fields, including geophysics (Shaw and Srivastava 2007; Fernandez-Alvarez *et al.* (2008); Fernandez-Martinez *et al.* (2008a,b).

PSO defines a swarm of particles (models) in an M-dimensional space. Each particle maintains memory of its previous best position $\mathbf{p}_i$ and velocity $\mathbf{v}_i$. At each iteration, a velocity adjustment of the particle is determined jointly by the previous best position occupied by the particle and the best position of the swarm. The new velocity is then used to compute a new position for the particle. The component of the velocity adjustment determined by the individual's previous best position

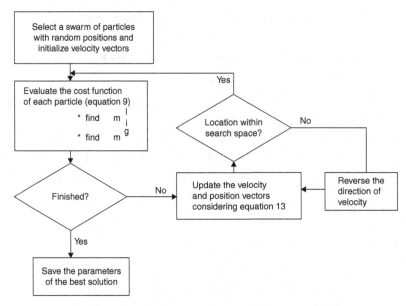

Figure 6.3 A flowchart for PSO. (From Shaw and Srivastava 2007.)

has been termed as the *cognition*, and the component influenced by the best in the population is the *social* part. The velocity update formula is given by

$$\mathbf{v}_i^k = \mathbf{v}_i^{k-1} + b.ran(\cdot)\left(\mathbf{m}_i^l - \mathbf{m}_i^k\right) + c.ran(\cdot)\left(\mathbf{m}_g - \mathbf{m}_i^k\right)$$
$$\mathbf{m}_i^{k+1} = \mathbf{m}_i^k + a\mathbf{v}_i^k,$$

(6.3)

where $\mathbf{m}_i^k$ is the current location, $\mathbf{v}_i^k$ is the current velocity, $\mathbf{m}_i^l$ is the best location so far, $\mathbf{m}_g$ is the best location achieved by the swarm prior to the kth iteration, ran($\cdot$) is a random number generator, and a, b, and c are constants. This algorithm is very intuitive and easy to program. A flowchart of the algorithm is given in Figure 6.3.

6.3 Simultaneous perturbation stochastic approximation (SPSA)

The SPSA method described in this chapter does not belong in the category of global optimization methods, but it involves a stochastic approximation of the gradient of the cost function needed in a local optimization method as described in Chapter 2. The method was originally proposed by Spall (1992, 2000) and has been applied to several optimization problems, including one on optimal well placement in reservoir engineering (Bangerth *et al.* 2006). Recall the model update formula for a simple gradient algorithm given by the following equation:

$$\mathbf{m}_{k+1} = \mathbf{m}_k - \alpha_k \mathbf{g}_k,$$

(6.4)

where

$$\mathbf{g}_k = \frac{\partial E}{\partial \mathbf{m}_k} = \left[\frac{\partial E}{\partial m_1} \frac{\partial E}{\partial m_2} \frac{\partial E}{\partial m_3} \cdots \right]^T, \tag{6.5}$$

is the gradient vector of the objective function with respect to model parameters, and α_k is step length. For many applications, it is not possible to evaluate this gradient vector analytically.

The brute-force approach to evaluating the gradient vector is to evaluate this by finite differencing. This requires that each model parameter be perturbed slightly while all other model parameters are kept fixed. With each new perturbed model, the forward problem is solved, and the new value of the objective function is evaluated. Using a forward finite-difference formula, we have

$$\frac{\partial E}{\partial m_j} = \frac{E(\mathbf{m} + \Delta m_j) - E(\mathbf{m})}{\Delta m_j}, \tag{6.6}$$

where Δm_j is a small perturbation of jth model parameter. Thus, the use of either a forward or backward differencing formula requires evaluation of an additional *NM* forward problem. Similarly, a central difference formula will require $2 \times NM$ forward model evaluations. For some problems, such as the 1D waveform inversion, one can take advantage of the forward modeling operator to reuse many of the forward modeling calculations in finite-difference computation (e.g., Sen and Roy 2003), but in general, this is not possible.

The SPSA offers an alternative approach to compute a stochastic approximation of the gradient that requires only two forward model evaluations. As the name SPSA implies, the algorithm simultaneously perturbs all the model parameters (unlike the approach described in the preceding paragraph) to stochastically approximate the gradient vector. In other words, each model parameter is perturbed at random to generate a random perturbation vector as follows:

$$\Delta \mathbf{m} = \left[\Delta m_1 \ \Delta m_2 \ \Delta m_3 \ldots \right]^T. \tag{6.7}$$

Now, two model vectors are generated using this model perturbation. They are given by

$$\begin{aligned} \mathbf{m}_1 &= \left[m_1 + c_1 \Delta m_1 \ m_2 + c_2 \Delta m_2 \ m_3 + c_3 \Delta m_3 \ldots \right]^T, \\ \mathbf{m}_2 &= \left[m_1 - c_1 \Delta m_1 \ m_2 - c_2 \Delta m_2 \ m_3 - c_3 \Delta m_3 \ldots \right]^T, \end{aligned} \tag{6.8}$$

where the perturbations to the model parameters are drawn from a distribution that satisfies certain conditions (Spall 2000). Sadegh and Spall (1996) showed that the best distribution for the components of the model perturbation vector is a symmetric Bernoulli distribution. An element of the gradient vector is approximated using the following central difference formula:

$$\frac{\partial E}{\partial m_j} = \frac{E(\mathbf{m}_1) - E(\mathbf{m}_2)}{2\Delta m_j}. \tag{6.9}$$

Note that in this formula, only two model vectors are used, for which objective functions are evaluated, and using these two objective function values, the entire gradient vector can be computed. The approach is simple: Spall (1992, 2000) provides some analysis on the validity of this approximation. However, its application to a wide range of optimization problems with large number of model parameters is still awaited.

7

Geophysical applications of simulated annealing and genetic algorithms

Many examples of iterative linear inversion of geophysical data have been reported, and these analysis methods are well understood. The limitations of these methods are also well known. Examples of the application of non-linear optimization in geophysics can also be found, but these are not as plentiful. Since the work of Rothman (1985, 1986), simulated annealing (SA) has been applied to a wide variety of purposes, such as coherency optimization for reflection problems (Landa *et al.* 1989), vertical seismic profile (VSP) inversion (Basu and Frazer 1990), dipmeter analysis (Schneider and Whitman 1990), seismic waveform inversion (Sen and Stoffa 1991), inversion of poststack data (Vestergaard and Mosegaard 1991), inversion of normal-incidence seismograms (Scales *et al.* 1992), reflection travel-time optimization (Pullammanappallil and Louie 1993), transmission tomography (Ammon and Vidale 1993), inversion of resistivity sounding data (Sen *et al.* 1993), magnetotelluric appraisal (Dosso and Oldenburg 1991), etc. Genetic algorithms (GAs) were introduced to geophysics by Stoffa and Sen (1991), which was closely followed by Sen and Stoffa (1992a), Sambridge and Drijkoniongen (1992), Scales *et al.* (1992), and Mallick (1999). The method has also been applied to earthquake location (Kennett and Sambridge 1992; Sambridge and Gallagher 1993) from teleseismic observations, migration velocity estimation (Jervis *et al.* 1993a; Jin and Madariaga 1993), and residual statics estimation (Wilson and Vasudevan 1991).

Since publication of the first edition of our book (Sen and Stoffa 1991), numerous applications of GAs and SA have been reported in the literature. Here we draw principally from our own work to illustrate how GAs and SA can be used in geophysics. We employ several seismic examples for the problem of 1D seismic waveform inversion and prestack migration velocity analysis. We also use examples from resistivity profiling and sounding, magnetotellurics (MT), and a problem in reservoir description. The examples are chosen to illustrate how SA and GAs work on synthetic data where we know what the answer should be and, in several cases, on real data to show that they are immediately applicable in the solution of real-world problems.

In the examples that follow, we illustrate how SA and GAs can be used to find a best-fitting model that explains the geophysical observations without the requirement of a good starting model. As discussed in Chapter 2, the minimum of an error function corresponds to the maximum of the posterior probability density function in many situations. Used in this way, these methods can also be called *maximum a posteriori* (MAP) *estimation algorithms*. In all the following examples we show how we can obtain an optimal model (answer) that corresponds to a solution that is likely to be at or near the global minimum of a suitably defined error function. At this stage we do not address the issues related to the uncertainties in the solution, although we realize that no description of the solution can be complete without describing its uniqueness and certainty. The methods of estimating those uncertainties in terms of posterior probability density functions (PPDs) and covariances will be described in Chapter 8.

In all the examples discussed herein, we are faced with many issues that are similar to those which must be addressed when employing any inverse method. Among these are incomplete data, e.g., limited-aperture seismic data, incomplete knowledge of the excitation function (the source wavelet in the case of seismic data), limited spatial and temporal bandwidth and hence resolution, calibration issues, and the validity and numerical accuracy of the modeling algorithm. Consequently, in some cases we employ nearly perfect data to show how the algorithm reaches a solution or to illustrate how various parameters affect the convergence rate and hence the computational cost of the algorithm.

Unlike gradient methods, GAs and SA do not require a good starting model. Instead, they only require knowledge of the limits of the search space that include the solution. Also, unless the gradient matrix or the Hessian required by gradient methods can be formulated analytically, leading to a subsequently rapid numerical evaluation, the cost of computing these derivatives numerically often makes SA and GAs a competitive alternative. Furthermore, unlike some of the gradient methods, SA and GAs do not require storage and numerical evaluation of the inverse of large matrices.

Another attractive feature of these algorithms is that they are generally applicable. Once the cost function has been defined for the problem of interest, any of the algorithms can be employed with equal ease. The cost function itself can be arbitrarily defined and even dynamically modified with minor implementation effort. These characteristics make experimentation with arbitrary measures of error, penalty functions and multi-objective optimization using mixed norms immediately possible.

Many of the algorithms, e.g., GAs, heat bath SA, and SA without rejected moves, also can be implemented on parallel computer architectures with minimal effort because multiple cost function evaluations are required per iteration. Aside from any natural parallelism that may be present in the problem, e.g., in seismology due to plane-wave and frequency decompositions, single nodes can be assigned the task of determining the cost function for a given model. This will further enhance the viability of many of the algorithms because parallel architectures have become more generally available.

7.1 1D seismic waveform inversion

In this problem, surface-reflection seismic data are analyzed to determine some combination of the compressional-wave velocity, shear-wave velocity, density, impedance, and Poisson's ratio. The earth structure is assumed to vary only with depth; i.e., the problem is 1D. Prestack seismic data are required because the change in the reflection response with the subsurface reflection angle is used to determine the physical properties of the medium. This change with angle affects not just the reflection coefficients but also the arrival times of the seismic energy.

Even if we assume a non-dissipative 1D isotropic elastic earth (i.e., anisotropy and attenuation are ignored), the problems inherent in this application include incomplete knowledge of the source wavelet and a limited spatial recording aperture. Further, even though our model of the earth is reasonably simple, a full reflectivity calculation that includes all compressional to shear-wave conversions and multiples is not a trivial numerical calculation. Since this calculation can be performed most rapidly in the frequency-horizontal wave-number ($\omega - k$) or frequency-horizontal ray-parameter ($\omega - p$) domain, it is preferable to transform the observed seismic data that are a function of source–receiver offset and travel time ($x - t$) to the domain of intercept time and ray parameter ($\tau - p$) by means of a plane-wave decomposition (Brysk and McCowan 1986). Because of the limited spatial aperture and discrete spatial sampling, the $\tau - p$ data that result from this transform will have coherent artifacts that should be minimized before the optimization proceeds. Wood (1993) addresses many of these issues before attempting iterative linear 1D seismic waveform inversion in the $\omega - p$ domain. Wood's proposed solutions (which include offset extrapolation, trace interpolation, calibration based on the seafloor multiple, and source wavelet estimation) can all be applied as part of the data preparation prior to the optimization.

In the examples that follow, we avoid many of the issues associated with real data by using nearly perfect synthetic data generated directly in the $\omega - p$ domain. Our purpose is to illustrate how the SA and GA algorithms work, not to address issues common to all inversion algorithms. Real-data examples are also presented, and one is compared with the iterative linear inversion of Wood *et al.* (1994). A second real-data example of a gas sand is also presented as an example of amplitude versus offset (AVO) analysis. In this example, the seismic modeling employed is overly simple (primaries only in the $x - t$ domain) but typical of many AVO implementations. Even so, the results obtained are reasonable.

It is important to remember that the forward modeling can be as simple or as complex as is warranted by the geologic situation. SA and GAs only require the evaluation of the cost function that measures the agreement (GAs) or disagreement (SA) of the synthetic data with the observed data. Further, in these algorithms, it is possible to gradually change the modeling or cost function as the algorithm proceeds. For example, it is possible to gradually change from an L_1 to an L_2 measure of error or to include linear inversion in a GA to speed convergence of the algorithm. Thus considerable flexibility exists in the modeling employed, the design of the cost function, and in designing hybrid combinations of linear and non-linear methods.

7.1.1 Application of heat bath SA

The first example is the application of heat bath SA to the problem of seismic waveform inversion from Sen and Stoffa (1991). The goal is to find the compressional-wave velocity, density, and impedance (product of velocity and density) for a 1D acoustic earth model consisting of ten layers shown by the solid lines in Figure 7.1a. The data to be matched are perfect in the passband 10 to 80 Hz and consist of nine $\tau - p$ seismograms from 0.0 to 0.4 s/km sampled every 0.05 s/km. The seismograms were generated for an acoustic earth model, and all interbed multiples but no surface multiples were included.

We now need to define our measure of misfit. For each trial of SA, a new 1D earth model is selected, and nine new seismograms are computed. These are compared with our target seismograms (Figure 7.1b). Sen and Stoffa (1991) used two definitions of error that correspond to the geometric and harmonic measures of misfit (Porsani *et al.* 1993). Since the seismograms are computed in the $\omega - p$ domain, these are given by

CC1:

$$E = -\frac{1}{N_p}\sum_1^{N_p} \frac{\sum_1^{N_\omega} u_o u_s^*}{\left(\sum_1^{N_\omega} u_o u_o^*\right)^{1/2} \left(\sum_1^{N_\omega} u_s u_s^*\right)^{1/2}}, \tag{7.1}$$

CC2:

$$E = -\frac{1}{N_p}\sum_1^{N_p}\frac{2\sum_1^{N_\omega}u_o u_s^*}{\left(\sum_1^{N_\omega}u_o u_o^*\right)^{1/2} + \left(\sum_1^{N_\omega}u_s u_s^*\right)^{1/2}}, \qquad (7.2)$$

where N_ω is the number of frequencies, N_p is the number of plane-wave seismograms, u_o are the observed data, and u_s are the current synthetic data being evaluated, and the asterisk represents complex conjugate. The real part of E is used, and we note that both measures of misfit correspond to the negative of a normalized correlation coefficient. Since CC1 is a geometric measure, it is not sensitive to any absolute amplitude differences. CC2 is sensitive to any scale-factor differences. In Figure 7.1 we used CC1 as the measure of misfit.

For heat bath SA, we need to define the starting temperature and the cooling schedule. In theory, we are required to start at a very high temperature, do many trials at each temperature, and then slowly lower the temperature. In practice, this is prohibitively expensive, and we must identify a parameter called the *critical temperature*. We employed an approach similar to Basu and Frazer (1990), who made several short SA runs at different fixed temperatures. Here we employed the cooling schedule suggested by Rothman (1986), $T_k = T_o(0.99)^k$, where k is the iteration number, and made several short SA runs beginning at different initial temperatures T_o. When the starting temperature is well above the critical temperature, the misfit is consistently high; i.e., the correlation is poor between the observed and synthetic data. When the starting temperature is well below the critical temperature, the algorithm becomes trapped in a local minimum if the initial model is far from the optimal solution. Cooling at too rapid a rate has a similar effect. Sen and Stoffa (1991) showed that by trying several initial temperatures, it is possible to bracket the critical temperature and then start the heat bath SA just above that temperature.

For this problem and using CC1, we found that by starting at $T_o = 0.005$ followed by slow cooling according to the schedule $T_k = T_o(0.99)^k$, many good solutions were quickly found, and the algorithm converged to near the global optimal solution within 100 iterations. The maximum correlation value obtained was 0.952. The results are shown in Figure 7.1, where the derived interval velocity, density, and corresponding impedances are shown. Also presented are the original and predicted seismograms and their differences. At this point, we could easily use our result as the starting model in an iterative linear inversion scheme to improve the correlation and further resolve the model parameters.

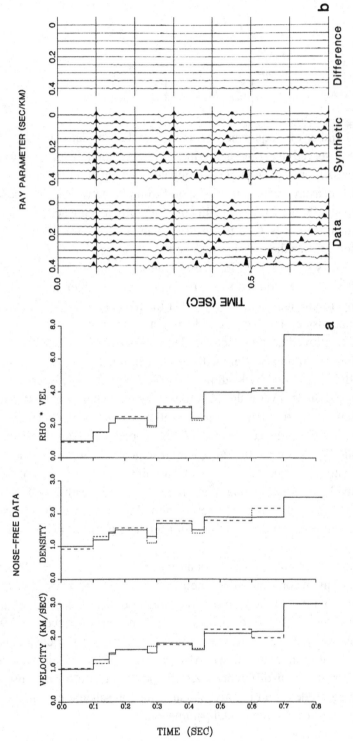

Figure 7.1 Example of seismic waveform inversion by SA: (a) Comparison between the true (*solid line*) and inverted (*dashed line*) profiles for a cross-correlation of 0.952. Note that the impedance match is almost exact. (b) Comparison between data and synthetic seismograms for the model shown by the dashed line. All the seismograms are plotted with the same gain.

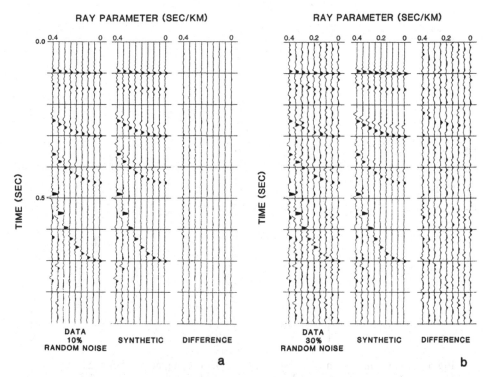

Figure 7.2 SA inversion of noisy synthetic data. Comparison between data and synthetic seismograms: (a) 10 percent random noise; (b) 30 percent random noise.

SA inversion is relatively robust in the presence of additive random noise. Figures 7.2 and 7.3 show the results obtained for an identical experiment but with white random noise that is 10 and 30 percent of the peak absolute value added to the seismograms. The predicted seismograms agree with the observations (Figure 7.2), and the derived models (Figure 7.3) are in reasonable agreement with the true structure but do degrade as the noise level increases.

Figure 7.4 shows the correlation values obtained at each iteration for the 10 and 30 percent additive noise cases. Note that in this example, only 80 iterations were used, and the maximum correlation values achieved were 0.823 and 0.506, respectively. These lower correlation values are a result of the noise and are expected for the amounts of noise present and the objective function used. The corresponding seismograms in Figure 7.2, however, are in very good agreement with the original data, even though the corresponding correlation values are lower than in the noise-free case (Figure 7.1).

In these examples, we constrained the thickness of the layers (in terms of their two-way travel time at normal incidence). This is a relatively severe constraint. In

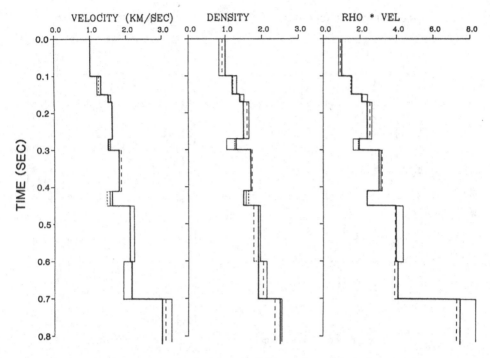

Figure 7.3 SA inversion of noisy synthetic data. Comparison between the true (*bold solid line*) and inverted velocity, density, and impedance profiles. The dashed line corresponds to the 10 percent random noise case, and the thin continuous line corresponds to the 30 percent random noise case.

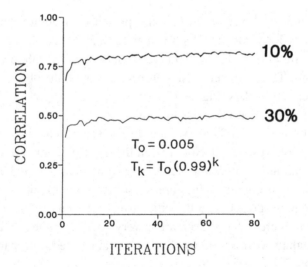

Figure 7.4 SA inversion of noisy synthetic data. Correlation versus iteration diagrams for 10 percent random noise (highest correlation = 0.823) and 30 percent random noise (highest correlation = 0.506). Note that the highest correlation value decreases with increasing random noise.

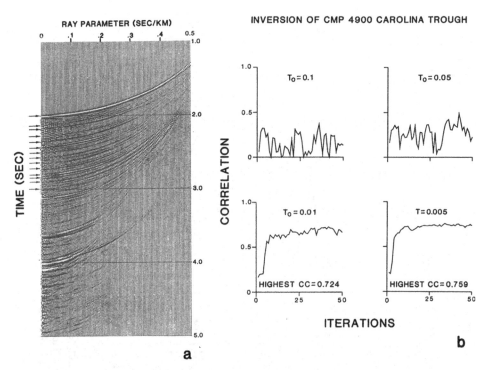

Figure 7.5 SA inversion of real data: (a) slant-stacked CMP gather that was used in the inversion; (b) plot of correlation versus iteration for different starting temperatures.

the real-data example below, we allow the thickness to be determined. In practice, for sparse models as in this example, we usually estimate the layer thicknesses and then allow the algorithm to search for the optimal values in the vicinity of these estimates.

Figure 7.5 shows a real common depth point (CDP) gather from the Carolina trough on the east coast of the United States after a plane-wave decomposition. Two-hundred and forty original $x - t$ seismograms for offsets up to 6.0 km every 0.025 km were used in the plane-wave decomposition. The fourteen reflection events identified for modeling are indicated by the arrows. In this example, the inversion is also used to define the optimal times within ±8 ms around these speci-fied times. Compressional-wave velocities between 1.0 and 3.0 km/s and densities between 1.0 to 3.0 g/cm³ were used in the search space. The source wavelet was estimated by windowing the seafloor reflection on the $p = 0.1$ s/km seismogram. Only 17 plane-wave seismograms were modeled between 0.1 to 0.302 s/km every 0.012 s/km (Figure 7.6). The correlation results per iteration for four different starting temperatures are shown in Figure 7.5b to the right of the plane-wave data. It is obvious from these plots that the critical temperature is below 0.05. Starting

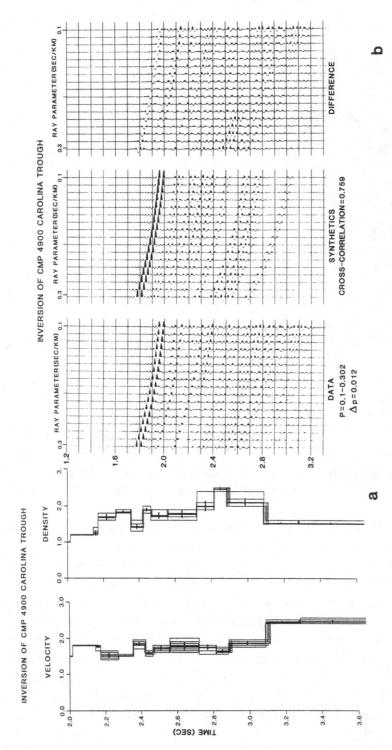

Figure 7.6 SA inversion of real data. (a) Velocity and density models with correlation values greater than 0.74. The bold line corresponds to the mean value, and the horizontal bars are the standard deviations. (b) Comparison between data and synthetic seismograms computed by using the model with a correlation value of 0.759.

with an initial temperature T_o of 0.01 or 0.005, followed by cooling using the same schedule as in our previous example, gave correlation values of 0.724 and 0.759, respectively, after only 50 iterations.

Figure 7.6 shows eleven models that gave a correlation value of 0.74 or greater. These were used to estimate the mean and variance. The bold lines correspond to the mean, and the horizontal bars indicate the standard deviations. This estimate is only based on a few good models. To obtain theoretically correct estimates of the mean and standard deviations, we need to use many more iterations and apply importance sampling techniques (see Chapter 8). Figure 7.6 also shows the estimated seismograms and their differences for the model with the maximum correlation of 0.759.

Since we do not know the earth structure for this real-data example, we evaluate the estimated interval velocity function for the model that achieved the maximum correlation value of 0.759. Figure 7.7 shows all the plane-wave data of Figure 7.5 after a $\tau - p$ normal move-out (NMO) correction (Stoffa *et al.* 1981). The result is excellent for the modeled seismograms up to $p = 0.302$. To the right we display the interval velocity function derived by SA and that obtained by interactively interpreting the $\tau - p$ travel times directly (Stoffa *et al.* 1992). For both the results, the corresponding average velocities are also shown and are nearly identical. Since the NMO is correct, we conclude that SA has performed as well as can be expected for the events selected (i.e., the model parameterization chosen) in estimating the interval velocities and event times.

7.1.2 Application of GAs

Our next example uses the GA described by Stoffa and Sen (1991) to solve a similar problem. Figure 7.8 shows nine ideal plane-wave seismograms for ray parameters 0.0 to 0.4 s/km sampled every 0.05 s/km. The bandwidth is again 8 to 80 Hz, as in our previous synthetic example. Our goal here is to illustrate how the GA works and how the algorithm's parameters affect the convergence of the algorithm. Figure 7.8 shows the nine reflection events identified for modeling, the actual 1D earth model, and the search space for velocity and density. These are the same as those used in the preceding example. GAs require a coding scheme to represent the model parameter values. Here we employed binary coding with a resolution in velocity of 0.02 km/s and 0.02 g/cm^3 for density.

In our initial experiments, we tried single- and multipoint crossover. We found that multipoint crossover worked slightly better in that it achieved a good solution in fewer generations. For this particular problem, this has some intuitive appeal because in a multipoint crossover scheme we are exchanging genetic information for the velocity and density within each layer.

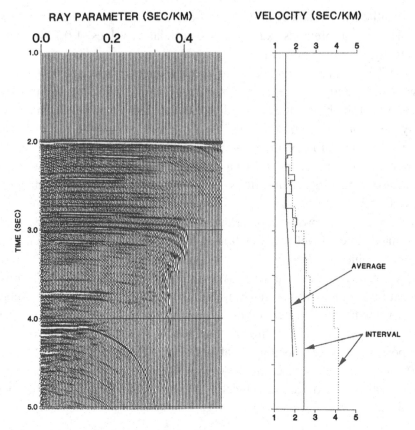

Figure 7.7 SA inversion of real data. NMO-corrected $\tau - p$ CMP gather; the velocity function obtained by SA was used to move out the data in the $\tau - p$ domain. The right-hand panel shows the velocity function used to move out the data and also a comparison between the average velocity obtained by SA and travel-time analysis. The solid lines correspond to the velocity functions derived from SA (correlation = 0.759), and the dotted line corresponds to the velocity function obtained by travel-time analysis.

The fitness is defined as the normalized cross-correlation defined by CC2. Unlike SA, in which we used the negative of a correlation function as the error function (Eqs 7.1 and 7.2), here we used the correlation function itself as the fitness function that the GA attempts to maximize. This harmonic measure of fitness is sensitive to scalar amplitude differences between the observed and synthetic data. For evaluation of the algorithm, we use 200 models and 200 generations. Initially, we fix the mutation probability p_m at 0.01 and vary the crossover probability p_x and update probability p_u.

Figure 7.9 shows the maximum and mean correlation values of the population at each generation for different combinations of p_x and p_u. The upper convergence

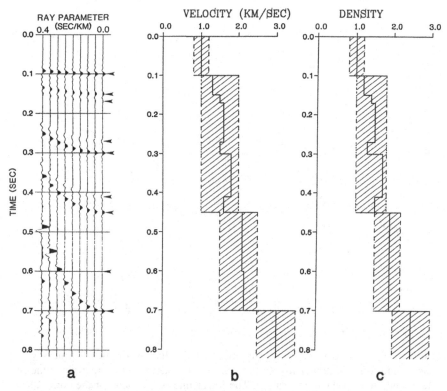

Figure 7.8 Example of seismic waveform inversion by a GA: plane wave, $\tau - p$, seismograms (*left*) used to evaluate GA optimization for 1D seismic waveform inversion. Seismograms correspond to the velocity and density models indicated by the solid lines. The seismic reflection events are indicated by the arrows, and the model search space used is indicated by the stippled areas.

plots have p_u fixed at 0.9, and p_x varies from 0.3 to 0.6 to 0.9. The higher the probability of crossover, the more extensively we will sample model space because more exchange of genetic information is likely to occur. This can be seen in the convergence plots because the algorithm converges more slowly; i.e., the mean fitness approaches the best fitness more slowly when p_x is 0.9 (*right*) compared with 0.6 (*center*) and 0.3 (*left*). Similarly, if no models from prior generations are allowed, $p_u = 0.0$, convergence is slow (Figure 7.9, *lower left*). By using a non-zero update probability, better models from the preceding generation are allowed back into the current population. When, for example, p_u is 1.0, if the preceding generation's model is better, it is always used. The result is rapid convergence (Figure 7.9, *lower right*). Using $p_u = 0.5$ (Figure 7.9, *lower center*) maintains diversity in the population with much better convergence than that achieved when no models from previous generations are included.

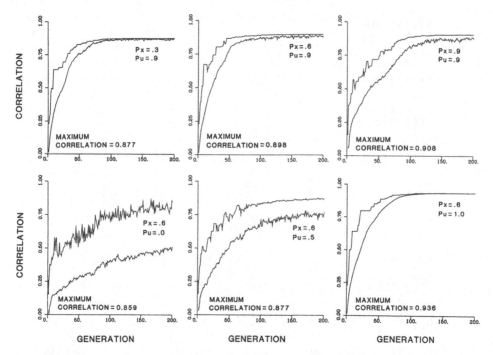

Figure 7.9 Seismic waveform inversion by GA: display of model fitness versus generation. The upper line in each plot is the maximum correlation value in the population for each generation. The lower line is the mean of the correlation values for each generation. When the lines are separated, the model population is diverse. When the lines are nearly coincident, the population has become homogeneous. Variations in the crossover probabilities p_x and the update probability p_u affect the rate of convergence of the population's mean correlation values to the maximum correlation. In this example, 200 models were evaluated every generation. The mutation probability was fixed at 0.01. When p_u is small, the mean of the population slowly converges to the maximum. The same effect is observed for a high crossover probability p_x.

Figure 7.9 shows that we can modify p_x and p_u to meet our expectations for the algorithm. If rapid convergence is desired, a low to moderate p_x, e.g., 0.3 to 0.6, can be specified. If slow convergence is desirable, the p_x should be high, e.g., 0.6 to 0.9. Similarly, if p_u is small, e.g., less than 0.5, convergence will be slower than when p_u is greater than 0.5. The designer of the algorithm can determine the goal of the optimization and the computational cost that is acceptable and define these parameters accordingly. It is usually desirable to search the model space reasonably well, which argues for moderate p_x and p_u values, subject, of course, to the computational resources available. If, however, we are interested in rapidly finding many good solutions to characterize uniqueness, p_x and p_u can be chosen accordingly, and the algorithm can be restarted many times using a new ensemble of randomly selected models for each restart (see Sen and Stoffa 1992a, and Chapter 8).

Increasing the mutation rate also increases the diversity in the population and slows convergence. In the example that follows, we used $p_x = 0.6$, $p_u = 0.9$, and $p_m = 0.01$. The number of models in the population also determines how well the model space is sampled and the rate of convergence. Stoffa and Sen (1991) show that for this problem, decreasing the number of models to 50 causes premature convergence, whereas using 400 models gives a good result but at considerable additional cost; e.g., 400 models for 200 generation is 80,000 forward calculations. Consequently, we use 200 models for 200 generations, i.e., 40,000 forward calculations.

Finally, we stretch the fitness by forming a temperature-dependent probability, as described in Chapter 5. Here we used an initial temperature of 1.5 followed by cooling with a decay rate of 0.99. This has the effect of de-emphasizing fit models prior to generation 42 and then exaggerating the differences between fit models after this generation.

Figure 7.10 shows the evolution of the solution using this algorithm. Four reflection events are monitored by displaying the fitness values at generations 1, 25, 50, 75, 100, and 200. For the first generation, the random sampling of the model space is obvious, with many high and low fitness values occurring. By generation 25, some localization has occurred, particularly for velocity (note the vertical alignment of the fitness for events 1 and 2). Also note that localization is occurring in a top-down fashion; i.e., the velocities of the upper layers are being determined before those of the deeper layers. By generation 100, the algorithm has essentially converged. The final generation, 200, shows that except for some random mutations, all the models have converged near to the actual solution, as indicated by the Xs.

The result of this run (the best-fit model) is shown in Figure 7.11. The derived velocity, density and corresponding impedances are shown as is the true model. The original plane-wave seismograms, those predicted by the final model, and their differences are shown to the right. The model result is reasonable, and the seismogram match is excellent, with the final correlation (fitness) value achieved being 0.992.

The exercise was repeated after the addition of random noise corresponding to 30 percent of the peak absolute value of the data. The result is shown in Figure 7.12. The result is again good, indicating that this GA (like the SA of the preceding example) is robust in the presence of random noise. The maximum correlation achieved in this case was 0.556, which is slightly higher than that obtained by heat bath SA.

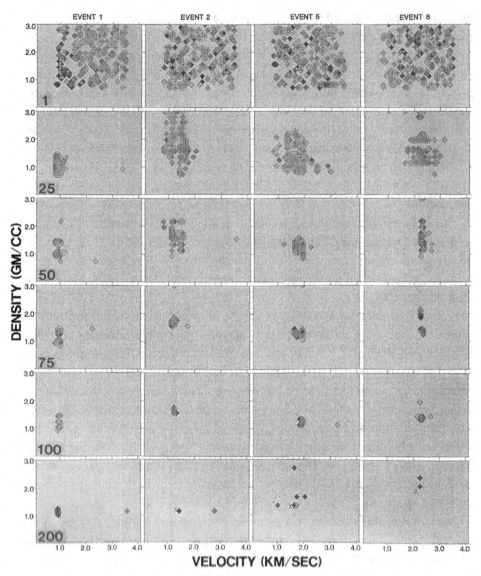

Figure 7.10 Seismic waveform inversion by GA: a contour display of the correlation values for four events from the 10-layer model evaluated in Figure 7.8, using 200 models at the end of generations 1, 25, 50, 75, 100, and 200. Each contour plot displays the correlation values as a function of velocity and density. Generation 1 shows the random distribution in model space used to start the process. After the twenty-fifth generation, the model parameters begin to localize, with the velocity being better defined in the upper two layers (events 1 and 2). As the process continues to the two-hundredth generation, the model parameter estimates converge toward the true model parameters, which are indicated by the X. Note that the velocity is determined better and more quickly than the density. The low correlation values away from the main clusters in the later generations are attributed to mutation.

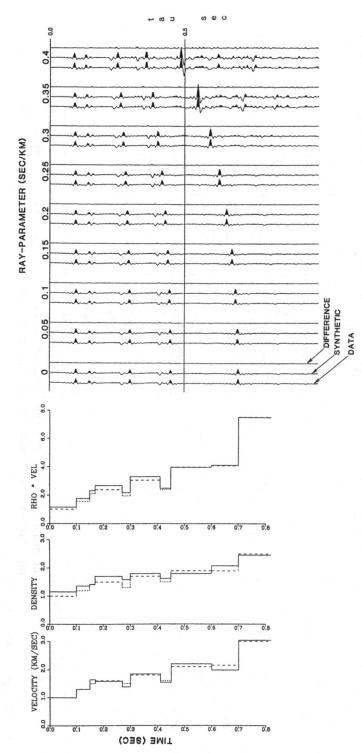

Figure 7.11 Seismic waveform inversion by GA. Comparison between the true model (*dashed line*) and the best model (*solid line*) corresponding to a correlation of 0.992 (see Figure 7.8) for noise-free data. Also shown are the data and synthetic seismograms calculated using the model shown by the solid line. Note that the match is excellent.

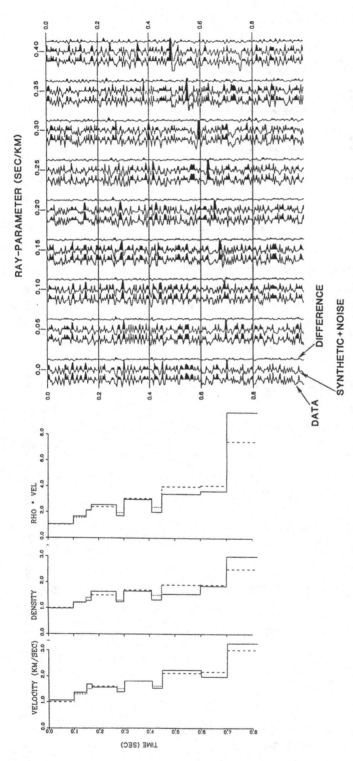

Figure 7.12 GA inversion of noisy synthetic data. The data of Figure 7.8 with 30 percent additive noise were used to evaluate the sensitivity of the GA to noise. The true (*dashed line*) model and the model (*solid*) corresponding to the maximum correlation of 0.556 are shown. Also shown are the original data with 30 percent additive noise and the synthetic seismograms calculated using the model shown by the solid line with the same noise. The differences between the two seismograms are small except for the high ray-parameter traces.

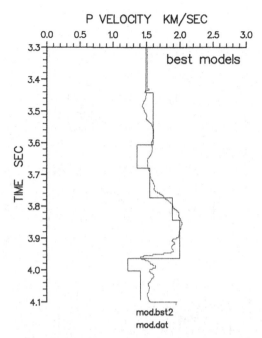

Figure 7.13 GA inversion of real data. GA result (*solid line*) for an eight-layer model where the layer thickness (in two-way normal time), compressional-wave velocity, and impedance were the model parameters. LI result (*dotted line*) from Wood (1993), where the background velocity was based on interactive $\tau - p$ NMO velocity analysis.

7.1.3 Real-data examples

Figure 7.13 compares iterative linear 1D seismic waveform inversion with a result obtained by non-linear optimization using a GA applied to a real common mid-point (CMP) gather shown in Figure 7.14a. The linear inversion was done in the $\omega - p$ domain by Wood (1993), and the background compressional-wave velocity was obtained by interactive analysis of the $\tau - p$ travel times and the $\tau - p$ NMO, as described by Stoffa *et al.* (1992). The interpreted velocity profile was then smoothed and used as the starting model in the linear inversion. For each two-way normal-time sample, the linear inversion algorithm estimated the compressional-wave velocity and impedance. The GA optimization was performed in the $x - t$ domain. 1D ray tracing was used to define the event arrival times and reflection coefficients and to correct for spherical divergence. The source wavelet was estimated by windowing the seafloor arrival. Eight reflection events were selected, and their approximate two-way normal times (for $x = 0$) were interpreted. The inversion then determined the actual layer-interval two-way times, compressional-wave velocity, and impedance. The results of the two inversion

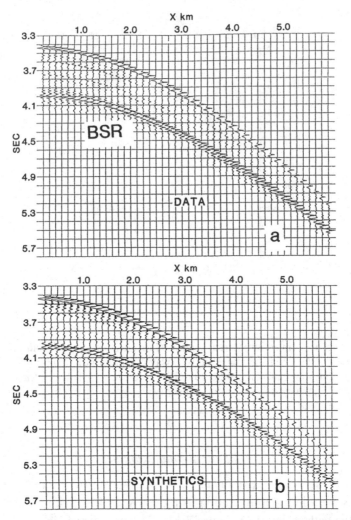

Figure 7.14 GA inversion of real data. (a) CMP used for GA inversion. Note the decrease in amplitude with offset of the seafloor reflection and the increase in amplitude with offset of the BSR event. The BSR is believed to be caused by gas hydrates. (b) Synthetics for the best-fit model obtained by GA inversion.

methods for compressional-wave velocity are shown on the same plot for comparison purposes in Figure 7.13. Overall, they are in good agreement considering the different parameterizations employed. Figure 7.14 compares the synthetic seismograms estimated for the best GA model with the original data. The agreement is very good, particularly for the major events, the seafloor reflection and the reflections at 3.9 to 4.0 seconds.

Our next example is an AVO study of conventional CMP data collected over the Cretaceous Colony Sand of southeastern Alberta (Hampson, 1991; Sen and

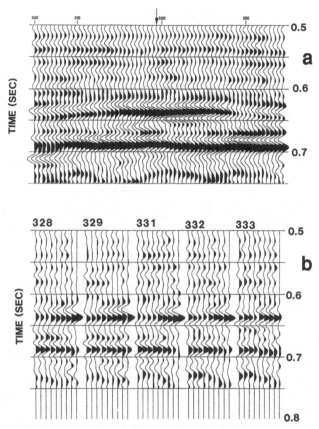

Figure 7.15 AVO inversion by GA. (a) A stack section of a seismic line collected over the Cretaceous Colony Sand of southeastern Alberta. Note the bright spot between CMPs 328 and 333 at a time of 0.63 seconds. (b) NMO-corrected CMP gathers showing increase in amplitude with offset. CMP 329 was used in the inversion; its location in the stack section is shown by the arrow.

Stoffa 1992b, 1992c). Figure 7.15a shows the stacked CMP data and Figure 7.15b the NMO-corrected CMP gathers. Note the increase of amplitude with offset for the event at 0.630 second. CDP 329 was modeled in $x - t$ using a GA (Figure 7.16). Note that the AVO anomaly is replicated in the synthetics calculated based on the best GA model. In this example, just the NMO-corrected CMP gather was modeled. Every time sample was considered an independent layer, but smoothing of the impedance and Poisson's ratio over three time samples was applied before calculating the synthetic seismograms. The resulting impedance and Poisson's ratio are shown in Figure 7.17. Note the decrease in Poisson's ratio that the GA best-fit model requires to model the seismic data.

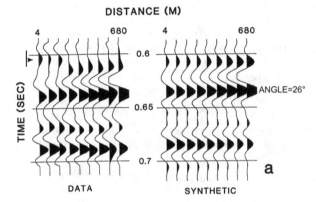

Figure 7.16 AVO inversion by GA. Comparison between the real data (CMP 329) and the corresponding synthetic seismograms for the best-fit model obtained by GA inversion.

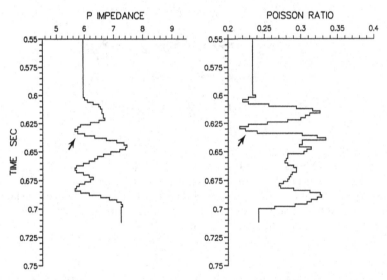

Figure 7.17 AVO inversion by GA. The best-fit model obtained from GA inversion of real data shown in Figure 7.16. The reflection from the gas sand has been modeled with a zone that shows a decrease in impedance and Poisson's ratio, as shown by the arrows.

7.1.4 Hybrid GA/LI inversion using different measures of fitness

Our next examples use our GA to illustrate the use of different measures of fitness and how linear inversion (LI) can be included at each generation to speed convergence and improve performance. In this synthetic example we use compressional-wave primaries only data but include loss due to shear conversion. Forty-one

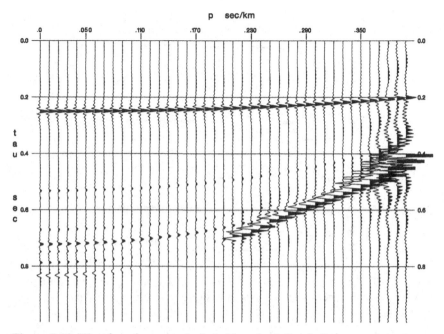

Figure 7.18 Waveform inversion by hybrid method. Synthetic $\tau - p$ seismograms used in the inversion employing GA and GA-LI. The seismograms were computed in the frequency range 8 to 80 Hz.

plane-wave seismograms are used from 0.0 to 0.40 s/km sampled every 0.01 s/km (Figure 7.18). The data are perfect in the frequency band 8 to 80 Hz. In this example the model is parameterized in terms of the compressional-wave velocity, impedance, and Poisson's ratio. The density and shear-wave velocity are derived quantities.

Following Porsani *et al.* (1993), we generalize our definitions for CC1 and CC2. CC1 is the geometric measure of fitness, and it can be generalized to

$$g_\alpha = 1 - \frac{\sum|y_i - x_i|^\alpha}{\sum|y_i + x_i|^\alpha \sum|y_i - x_i|^\alpha}. \tag{7.3}$$

CC2 is the harmonic measure of fitness, and the general form is

$$h_\alpha = 1 - \frac{2\sum|y_i - x_i|^\alpha}{\sum|y_i + x_i|^\alpha + \sum|y_i - x_i|^\alpha}, \tag{7.4}$$

where x_i is the observed data and y_i the synthetic data, respectively.

When α is 2, we have the original L_2 definition of fitness. If α is 1, the L_1 fitness measure is defined. When α is less than 1, a fractional measure of fitness is

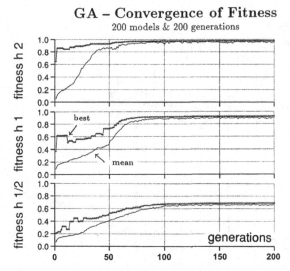

Figure 7.19 Waveform inversion by hybrid method. Convergence plots for a pure GA for the h_2, h_1, and $h_{1/2}$ fitness. Note the slower convergence of the $h_{1/2}$ fitness.

defined that de-emphasizes differences in the observed data and the calculated synthetic seismograms. As we will see below, this apparently has some advantage for the GA.

Figure 7.19 shows the convergence of fitness for the harmonic fitness measures h_2, h_1, and $h_{1/2}$. We can see from the plots that each measure of fitness converges to a different maximum value because of the way each measure is defined. Note, however, the way that convergence proceeds. In the h_2 case, the population becomes homogeneous faster (mean fitness converges to the maximum fitness by generation 50) than either the h_1 or $h_{1/2}$ examples. In this example, the best models for the h_1 and $h_{1/2}$ measures were closer to the true model than that obtained for the h_2 measure. We attribute the improved performance to the de-emphasis of differences in the observed and computed data that occurs with h_1 and $h_{1/2}$, which gives the algorithm more opportunity to search the model space. Since the mean fitness of the population converges toward the best fitness more slowly, diversity in the population is being maintained for a longer time, allowing the model space to be searched more thoroughly.

Next, for the same problem, we include one step of an LI in each generation of the GA. We select the best model of each generation and use it as the starting model. We take our fitness measure and define a measure of error $e = 1. -\text{fitness}$. Then, following Porsani *et al.* (1993), we define an update using

$$\mathbf{m}_{k+1} = \mathbf{m}_k + \left[\mathbf{G}_k^T \, \mathbf{F}_k^2 \, \mathbf{G}_k\right]^{-1} \mathbf{G}_k^T \, \mathbf{F}_k \, \mathbf{f}_k, \tag{7.5}$$

where F_{jj} are the non-zero diagonal elements of $\mathbf{F}_k^2$ given by

$$F_{jj} = 4\beta^2 \left[\Delta d \left(\mathbf{m}_k, t_j \right)^2 \right]^{2\beta-2} \Delta d \left(\mathbf{m}_k, t_j \right), \tag{7.6}$$

and the element f_j of the vector that results from the product $\mathbf{F}_k \mathbf{f}_k$ is equal to

$$f_j = 2\beta \left[\Delta d \left(\mathbf{m}_0, t_j \right)^2 \right]^{2\beta-1} \Delta d \left(\mathbf{m}_0, t_j \right). \tag{7.7}$$

For the L_2 case, $\beta = \frac{1}{2}$, $\mathbf{f}_k$ becomes the error vector between the observed and calculated data, $\mathbf{f}_k = \boldsymbol{d}_{\text{obs}} - g(\mathbf{m}_k)$, and the diagonal elements of the matrix $\mathbf{F}_k$ are equal to unity, $F_{jj} = \text{sgn}\left[\boldsymbol{d}_{\text{obs}}^j - {}^j_{\mathscr{g}} \left(\mathbf{m}_k \right) \right]$ such that $\mathbf{F}_k^2$ is an identity matrix implying that the preceding equation is the usual least squares solution for the L_2 norm:

$$f_j = 2\beta \left[\Delta d \left(\mathbf{m}_0, t_j \right)^2 \right]^{2\beta-1} \Delta d \left(\mathbf{m}_0, t_j \right). \tag{7.8}$$

The new LI model simply replaces the original best model in the population, and the GA continues. The idea is that the best model of each generation should be improved by the LI step, and the GA should converge to a better solution and do so more rapidly. This is in fact what happens.

The preceding example was repeated for each of the three measures of fitness. Plots are shown in Figure 7.20, which compares the best fitness from the previous example with the best fitness when one LI step is included. The GA with LI quickly moves to better fitness values compared with just the GA. Also note that the values obtained persist for several generations, and then large improvements occur. We interpret these plateaus as local fitness maxima that the LI continues to find because the best GA model has not changed sufficiently. When the GA finds a best model that is sufficiently different, the LI moves it to a new local fitness maxima that represents a significant improvement. Each measure of fitness performs in a similar way, and h_1 and $h_{1/2}$ achieve a better final fitness value than the preceding example, where LI was not included.

The resulting models are shown in Figure 7.21. The h_2 result is not as good as either the h_1 or $h_{1/2}$ results. Both the h_1 and $h_{1/2}$ results are excellent, and for this perfect-data example, even Poisson's ratio is being estimated reasonably well. This exercise is an example of how different measures of fitness can easily be employed and how LI can be incorporated into the algorithm. The actual performance achieved will depend on the signal-to-noise ratio and the type of noise present in the data, but experimentation to determine if, for example, the h_2 or h_1 measure should be employed is a straightforward task using a GA or SA.

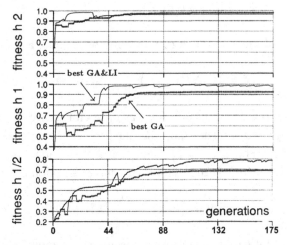

GA and GA&LI – Best Fitness
200 models & 175 generations

Figure 7.20 Waveform inversion by hybrid method. Convergence plots for the best models for the pure GA and the GA plus LI algorithms for h_2, h_1, and $h_{1/2}$ fitness. Note the improved performance of the h_1 and $h_{1/2}$ fitness compared with the pure GA.

7.1.5 Hybrid VFSA inversion using different strategies

It is not as obvious how LI can be added to very fast simulated annealing (VFSA) to improve performance. The conjecture is that as VFSA samples the model space and finds good models, perhaps we should look closer in the vicinity of these good models to see if a better solution exists in this part of the model space. One obvious option is to do an LI at the end of the VFSA run using the best model found as the starting point. If this can be done easily for a particular problem, it is an obvious way to try to improve the final result. During the VFSA run, however, it is not clear which strategy to employ. If we pursue a gradient-descent algorithm too early or too frequently, we may find a good but suboptimal solution. There is a tradeoff between the amount of sampling we will allow and the pursuit of good but possibly locally optimal solutions.

Chunduru *et al.* (1997) explored several strategies that incorporate LI into VFSA and called these *hybrid optimization methods*. They evaluated their strategies for two very different geophysical problems: the inversion of resistivity sounding data and seismic velocity analysis. They kept track of the number of function evaluations that were required for each strategy they tested and compared the results obtained. The details can be found in their study. But here we summarize their findings. The idea to employ LI when VFSA finds a new and significantly better

GA & LI – Model Comparisons

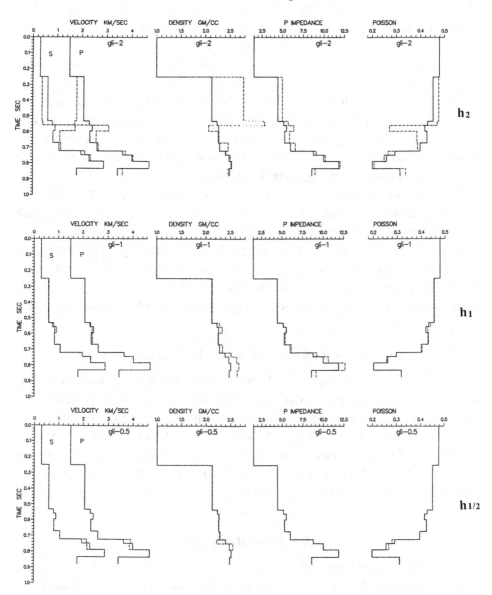

Figure 7.21 Waveform inversion by hybrid method. Model parameter results for the GA with LI run using the h_2 (*above*), h_1 (*center*), and $h_{1/2}$ (*below*) fitness. The results from the h_1 and $h_{1/2}$ fitness examples are the best and significantly better than the h_2 fitness.

model than previously found does speed convergence. There is a tradeoff between what is considered significantly better and the rate of convergence and the likelihood of getting trapped in a local minimum. They found that by requiring the current error to be less than the initial error by a specified percentage and by also requiring that this error be less than the best error found so far by another percentage before applying LI was a good strategy. In this way, the hybrid algorithm would employ VFSA during the early trials, widely sampling the model space, and perform LI primarily during the later trials, where VFSA is predominately finding good models. This strategy proved efficient in that the number of function evaluations was reduced. The percentages to use are likely problem-dependent and need to be ascertained as part of algorithm development.

In addition to the changes in absolute error, Chunduru *et al.* (1997) evaluated hybrid algorithms that employed LI based on the changes in the error gradient with iteration. This strategy also proved effective. Combinations of the two approaches are obviously possible. VFSA can be modified in numerous ways to generate problem-specific hybrid algorithms. Care must be taken to ensure that adequate sampling of the model space is performed and that the LI algorithm does not dominate the hybrid strategy until VFSA is searching in the vicinity of the global solution.

7.2 Prestack migration velocity estimation

Our previous seismic examples used reflection travel times and amplitude variations to estimate compressional- and shear-wave velocities, impedance, and Poisson's ratio. In these examples, the earth was assumed to vary only with depth. Here we will illustrate the use of GAs and VFSA to derive estimates of just the compressional-wave velocity but for laterally varying media. First, however, we will go back to the 1D earth problem and illustrate two different objective functions and different ways to parameterize the subsurface.

7.2.1 1D earth structure

Our first objective function is the migration misfit criterion (MMC) (Jervis *et al.* 1993a). The idea behind this definition of misfit is that if one or more nearby or adjacent shot records are migrated to depth, they will have sampled many of the same parts of the subsurface. A comparison of the individual images should agree to the maximum extent possible when the velocity function used to generate the images is correct. If the earth is 1D, the images from adjacent gathers should be identical.

Our first example illustrates how to apply this principle to the problem of determining interval velocities from $\tau - p$ data using the idea of $\tau - p$ NMO

(Stoffa *et al.* 1981). To do this optimization, the objective function is defined by evaluating the difference between the $p = 0.0$ s/km seismogram and all the other seismograms after the NMO correction. Only when all the other seismograms have been NMO-corrected properly will they be in phase with the target $p = 0.0$ s/km seismogram. The measure of agreement used was the h_1 measure of fitness described earlier in Eq. (7.4), where the y_i are now the $\tau - p$ NMO-corrected seismograms for all the available ray parameters at all two-way normal times. In this problem, the x_i are the samples of the $p = 0.0$ s/km seismogram that are compared with each NMO-corrected $p \neq 0$ seismogram. To minimize amplitude variations between the seismograms, the square root of each data sample (with polarity preserved) was used rather than the original data values, and each of the resulting seismograms was normalized to have a maximum absolute value of 1. The synthetic data (to be used as observations; shown in Figure 7.22) were generated for an 84-layer 1D earth model based on a blocked well log shown by the solid line in the upper right panel in Figure 7.23.

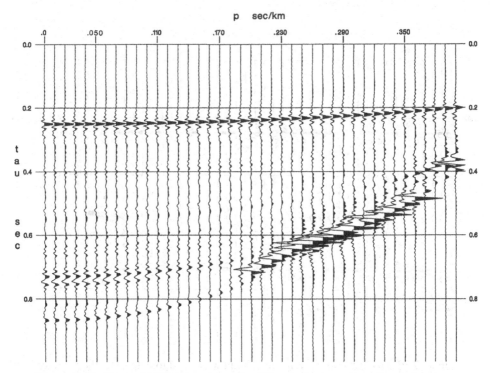

Figure 7.22 GA and VFSA inversion for 1D velocity structure. Synthetic $\tau - p$ seismograms used in the GA and VFSA inversion employing migration misfit and reflection tomography criteria. The seismograms were computed in the frequency range 8 to 80 Hz. Note the postcritical reflection events for the high-ray-parameter regions.

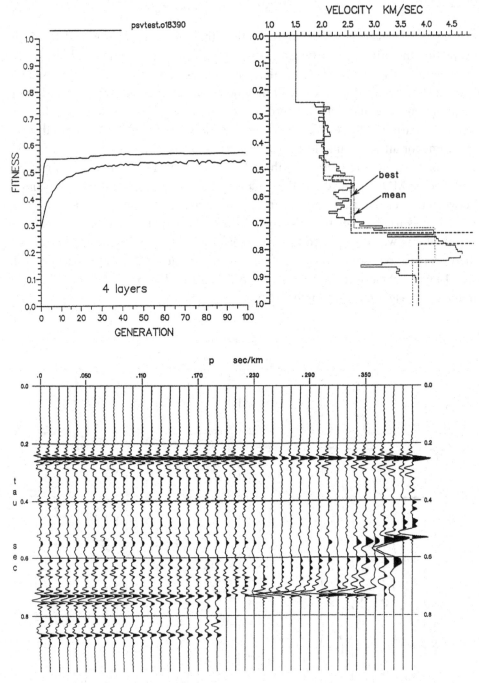

Figure 7.23 GA and migration misfit criteria were used to determine the velocity function for $\tau - p$ NMO corrections. The $p = 0.0$ s/km seismogram was the target seismogram, and four constant-velocity layers below the first layer were used to define the velocity model. The upper left plot shows the best and mean fitness values. Convergence is achieved by generation 30, even though the algorithm continues to generation 100. The original 84-layer velocity model and the estimated best and mean velocity inversion results are shown in the upper right. The $\tau - p$ NMO corrected seismograms are shown at the bottom. Each seismogram is individually scaled to its peak absolute value for display purposes.

We also illustrate how the model parameterization affects the results of the inversion. Figure 7.23 shows the results of a GA run using MMC for 200 models and 100 generations. In this example, the 1D earth model beneath the first layer was approximated by four constant-velocity layers. The GA that searched for velocities in the window 1.5 to 5.0 km/s found the best and mean layer thicknesses and velocities that can be compared with the true velocity function (Figure 7.23). The optimization has been successful in finding a reasonable estimate of the average velocity within each of the intervals it has selected. The $\tau - p$ NMO-corrected data are also displayed in Figure 7.23 to evaluate the results of the algorithm. All the reflection events are nearly flat, indicating that the velocity model was adequate.

Next, the same exercise was carried out using a different model parameterization scheme. The earth model below the first layer was parameterized using four spline nodes. Both the value and the position in depth of the spline nodes were determined by the inversion. Figure 7.24 shows the results. Except for the post-critical events, the results are nearly identical, even though the velocity models are different, indicating for these fitness criteria the degree of sensitivity to NMO for these data. In both these examples, the fitness versus generation plots indicate that convergence was achieved by generation 30. Thus, while 20,000 NMO corrections were evaluated to achieve the results shown, 6,000 probably would have been adequate.

Figure 7.25 is a similar example where the misfit criterion is the same, but nine spline nodes were used to describe the velocity model beneath the first layer. In this example, VFSA was used to do the optimization. Using this optimization algorithm, only 500 NMO corrections were required: 100 temperatures with five evaluations per temperature. The original temperature was 1.5 with a decay rate of 0.9999. The temperature schedules for the model parameters and acceptance criteria were the same. The decrease in misfit is shown in the upper left of Figure 7.25. Using nine spline nodes, we obtained a better result than in either of the two previous examples because the best model better approximates the slowly varying parts of the true velocity function. Once again, the precritical events are properly corrected for their $\tau - p$ NMO, and the main differences occur principally for the postcritical events (see the lower right of Figure 7.25 compared with Figure 7.24).

We now change the definition of misfit to what we will term *reflection tomography*. The basic idea of this misfit criterion is to predict the arrival times of all the reflections in the original data without having to *pick* these times manually. If an estimate of the subsurface reflectivity is available, this can be used to generate seismic arrivals to simulate a shot record or for a 1D earth model the corresponding $\tau - p$ data. The amplitude variations are not of interest, just the reflection travel (or vertical delay) times.

For our 1D examples, we use the $p = 0.0$ s/km seismogram as an estimate of the reflectivity. All the other seismograms are predicted using this reflectivity

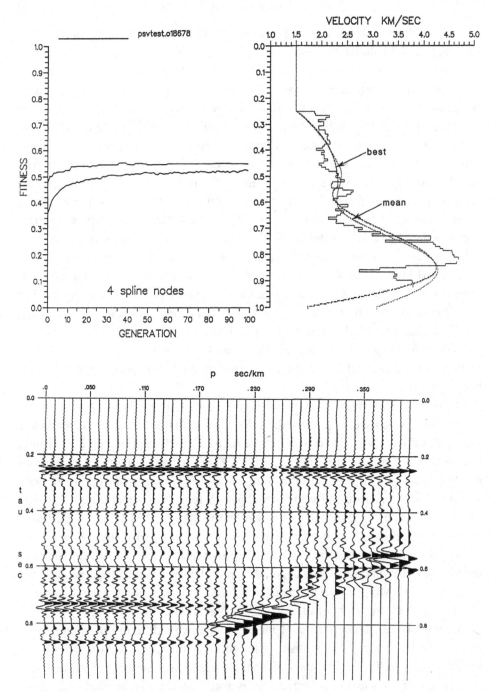

Figure 7.24 GA and migration misfit criteria were used to determine the velocity function for $\tau - p$ NMO corrections. The $p = 0.0$ s/km seismogram was the target seismogram, and four spline coefficients below the first layer were used to define the velocity model. The upper left plot shows the best and mean fitness values. The best and mean velocity inversion results are shown in the upper right. The $\tau - p$ NMO-corrected seismograms are shown at the bottom. Each seismogram is individually scaled to its peak absolute value for display purposes.

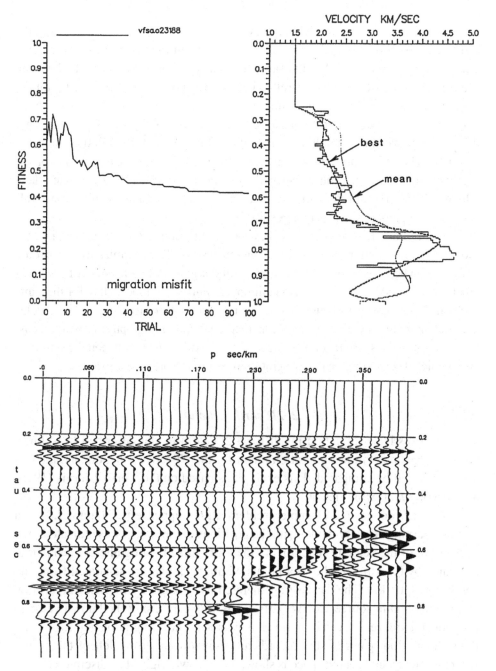

Figure 7.25 VFSA and migration misfit criteria were used to determine the velocity function for $\tau - p$ NMO corrections. The $p = 0.0$ s/km seismogram was the target seismogram, and nine spline coefficients below the first layer were used to define the velocity model. The upper left plot shows the decrease in the misfit for 100 temperatures. Five evaluations per temperature were used. The best and mean velocity inversion results are shown in the upper right, which better approximates the slowly varying components of the original velocity than in Figures 7.23 and 7.24. The $\tau - p$ NMO-corrected seismograms are shown at the bottom. Each seismogram is individually scaled to its peak absolute value for display purposes.

estimate and the velocity model being evaluated. Before the comparison using the h_1 measure of correlation, the amplitude variations in the observed and predicted data were again minimized in the same manner as in the MMC described earlier. Figure 7.26 shows the result for reflection tomography. VFSA was used with the same temperature schedule and number of function evaluations as before. Nine spline nodes were also used to describe the model below the first layer. The only difference between this result and that obtained in Figure 7.25 is in the definition of misfit. The reflection tomography criterion result is better, particularly for the high-velocity zone. This can be verified by inspection of the NMO-corrected data, where the postcritical reflection events form a nearly continuous suite of arrivals not present in any of the previous results.

The preceding 1D examples show how our description of the subsurface, i.e., the model parameterization, affects the inversion results. The optimization procedure employed, whether GA or SA, will always find a good solution subject to the parameterization and the misfit criterion employed. Further, it appears that for the same problem, VFSA performs better than a GA; i.e., convergence to a very good model is achieved more rapidly. Finally, it appears that reflection tomography is a better measure of misfit in the sense that it is more sensitive to differences between the observed and predicted seismograms than migration misfit in the preceding examples.

7.2.2 2D earth structure

We now investigate migration misfit for a 2D earth model described by splines. Our first example will use migration misfit for the objective function, and the GA and VFSA will be compared. This example is based on the work of Jervis (1993) and Jervis *et al.* (1993a, 1993b).

Figure 7.27 defines the example problem. In the upper left of Figure 7.27 is a laterally varying velocity model with the positions of the shot points indicated. Just to the right of the velocity function are the four pairs of shot gathers. Each pair is separated by 0.5 km, and the shot points within each pair are separated from one another by 0.05 km. The data were generated using a finite-difference algorithm, and the maximum source-receiver offset was 1.75 km. Using the known velocity function, the image that can be formed using a Kirchhoff migration algorithm, and the available four pairs of shot records are shown in the lower left of the figure, whereas the velocity search space is shown in the lower right. The discrete velocity values correspond to the possible values in the GA. The true image of Figure 7.27 is the desired result.

The true 2D velocity model was approximated by 30 spline nodes. Figure 7.28 shows the results obtained using VFSA and a GA. Sixty models for 20 generations were used in the GA, whereas VFSA used 300 temperatures and one evaluation per temperature.

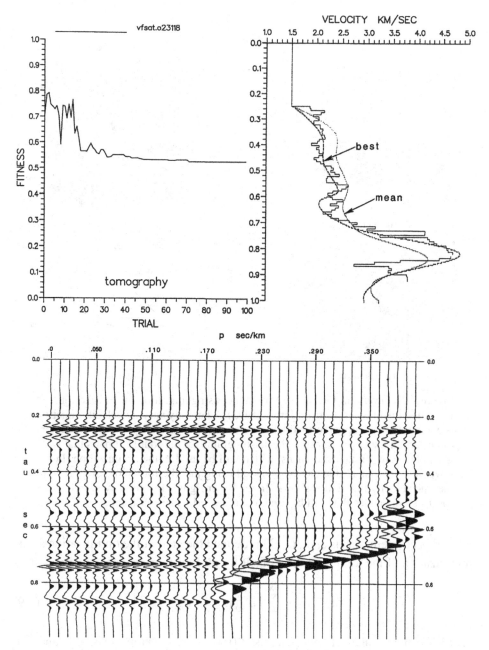

Figure 7.26 VFSA and reflection tomography criteria were used to determine the velocity function for $\tau - p$ NMO corrections. The $p = 0.0$ s/km seismogram was used for the reflectivity estimate, and nine spline coefficients below the first layer were used to define the velocity model. The upper left plot shows the decrease in the misfit for 100 temperatures. Five evaluations per temperature were used. The best and mean velocity inversion results are shown in the upper right and approximate the slowly varying components of the original velocity function very well, particularly for the high-velocity zone near the base of the model. The $\tau - p$ NMO-corrected seismograms are shown at the bottom of the figure. Each seismogram is individually scaled to its peak absolute value for display purposes. Note in this result the continuity of the postcritical reflections at the end of the seismograms for the high-ray, e.g., greater than 0.18 s/km, parameters.

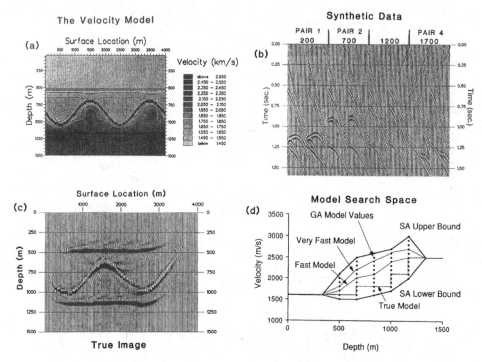

Figure 7.27 Inversion for 2D velocity model using migration misfit criterion. 2D migration misfit example from Jervis (1993). A laterally varying velocity function is contoured in the upper left with the source positions of four pairs of shot gathers indicated. In the upper right are the seismograms for the four shot pairs that were generated using finite differences, which will be used in the optimization. In the lower left, the image that can be obtained using just the four shot pairs but with the true velocity function is shown. In the lower right is the velocity search space for the spline nodes that will be used to parameterize the subsurface.

For this problem, Jervis *et al.* (1993a) defined the goal of the optimization as the minimization of the total migration misfit for all four migrated shot-gather pairs:

$$\text{Misfit}_{\text{MMC}} = \frac{1}{N} \sum_{k=1}^{N} \left[d_{\text{mig}}^{2,k}(x,z) - d_{\text{mig}}^{1,k}(x,z) \right]^2, \tag{7.9}$$

where N is the number of shot pairs used (four in this example) and $d_{1,k}^{mig}$ and $d_{2,k}^{mig}$ correspond to the first and second depth-migrated shot gathers for the kth shot pair, respectively. Using MMC, wavelet estimation and the coupling between reflectivity and velocity, which are common problems in approaches based on forward modeling, are no longer significant issues.

The migrated image using the initial model for VFSA is shown in the upper left of Figure 7.28, and the migrated image using the best model of the first generation of the GA is shown in the upper right. The best GA model of the first generation

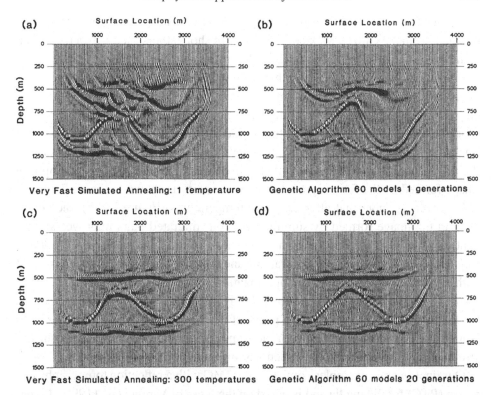

Figure 7.28 Inversion for 2D velocity model using migration misfit criterion. Images that correspond to VFSA and GA optimization of migration misfit between the shot pairs in Figure 7.27. The upper images are for the first VFSA model (*left*) and for the best model (*right*) of the 60 used in the GA for the first generation. The lower figures show the final images that correspond to the VFSA optimization (*left*) and the GA (*right*). VFSA has produced a very good result after only 300 migration misfit evaluations, whereas the GA result required 1,200 evaluations, i.e., 60 models for 20 generations (from Jervis 1993).

is (of course) better than the randomly selected model that initiates VFSA. By the end of the optimization, however, VFSA has converged to a somewhat better solution. Interestingly, the GA result is better for the second sinusoidal reflector, whereas the VFSA result is better for the deepest horizontal reflector (compare, e.g., Figures 7.27 and 7.28).

In this example, only 300 objective function evaluations were made for VFSA, whereas 1,200 were required for the GA. Each objective function evaluation required the Kirchhoff migration of eight shot gathers and the calculation of the migration misfit between each of the four shot-gather pairs. VFSA has produced a very acceptable result for one-fourth the computational effort. This result could now be used as the starting velocity model in an iterative linear inversion scheme, if desired.

Our next example is also from Jervis (1993) and illustrates using the MMC with real data from the Gulf of Mexico. Four pairs of shot gathers each with a maximum source-receiver offset of 2.31 km sampled every 0.07 km were used in a VFSA optimization. The shot gathers within each pair were separated by 0.14 km, and the pairs were separated by 1.05 km. To minimize the number of inversion parameters, spline nodes were spaced horizontally at every 0.962 km, although they were spaced every 0.1875 km vertically to provide better vertical resolution. Two of the images obtained during the optimization are shown in Figure 7.29. Remember, it is the migration misfit between pairs that is used in the inversion, not the composite image formed by the four migrated shot-gather pairs as shown in Figure 7.29.

However, we note that after only 100 temperatures (Figure 7.29b), the coherency in the image formed by summing the four migrated pairs improves significantly, indicating that the migration misfit must have decreased. The final velocity function was obtained after 300 iterations and then used to migrate every third shot gather between those used in the optimization so that better subsurface coverage could be obtained. The result is shown in Figure 7.30 (*lower right*), which should be compared with the intermediate result of Figure 7.29b. Figure 7.30 (*lower left*) is a second real-data example from the same line with more complex subsurface structure that was analyzed in an identical manner.

Our final seismic examples compare the MMC with a reflection tomography criterion (RTC) for 2D media and is based on the work of Varela *et al.* (1994). Figure 7.31 shows a velocity model that varies vertically and horizontally. Velocities vary from 2.2 to 4.7 km/s. Note the high-velocity layers enclosed in the syncline between the lower-velocity horizontal layers. For purposes of inversion, this 2D model will again be parameterized by splines. The positions of the ten vertical and five horizontal spline nodes are indicated by the black dots in Figure 7.31.

Three pairs of shot gathers and the zero-offset seismic section were generated using the split-step Fourier algorithm (Stoffa *et al.* 1990a, 1990b). The positions of the shot gathers are indicated by the asterisks along the surface of the velocity model. Each shot gather is separated by 0.09 km, and the shot pairs are separated by 0.45 km. The maximum source–receiver offset is 3.21 km and is located to the right of each source position. The image that can be obtained by migrating the six shot gathers with the known velocity model is shown in the lower half of Figure 7.31. This is the *target* for the optimization.

The first shot gather of each pair is shown in Figure 7.32 after application of automatic gain control (AGC) and an external mute. Below the shot gathers, the zero-offset section is displayed. VFSA will be used to minimize the total of the migration misfit of the three migrated shot pairs just as in our previous example. RTC will use the depth migration of the zero-offset section as a reflectivity estimate and then compute synthetic shot gathers for the first shot of each pair. The

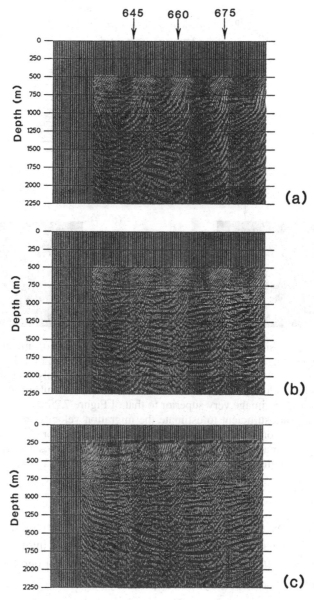

Figure 7.29 Inversion for 2D velocity model using migration misfit criterion. Improvement in the migrated images obtained using VFSA for a real-data example. In this case, four shot-gather pairs were used to form the migration misfit: (a) initial image; (b) after 50 iterations; (c) after 100 iterations, it finds better solutions by minimizing the migration misfit. The coherence in the corresponding image formed by summing the four migrated shot gathers improves. (From Jervis 1993.)

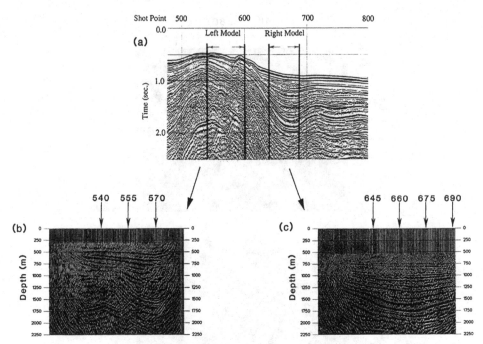

Figure 7.30 Inversion for 2D velocity model using migration misfit criterion. (*Lower right*) Image formed using every third shot gather and the final velocity function found by VFSA after 300 iterations and based on only four shot-gather pairs. The final velocity function combined with improved subsurface coverage makes the resulting image very superior to that of Figure 7.29(b). But using only four pairs appears sufficient to estimate the migration velocity in this example. (*Lower left*) A second real-data example of prestack migration velocity analysis and subsequent image formation. Both prestack migration results are from the areas indicated on the stacked data line in the upper part of the figure. (From Jervis 1993.)

amplitudes of the observed and predicted shot gathers will be equalized using AGC because it is only the agreement of the travel times that is of interest. All comparisons are in the 0- to 40-Hz passband. External mutes are applied before the comparisons are made. The RTC error is computed by

$$
\text{Misfit}_{\text{RTC}} = \frac{1}{N} \sum_{k=1}^{N} \left[d_{\text{obs}}^k (x,t) - d_{\text{syn}}^k (x,t) \right]^2,
\tag{7.10}
$$

where N is the number of shot records used, and $d_{obs}{}^k$ and $d_{syn}{}^k$ are observed and synthetic data after amplitude equalization of the kth shot gather, respectively.

Figure 7.33 shows the decrease in error for MMC and RTC. Note that RTC reaches a lower value. Both methods used the same initial temperatures, 1.0, for the acceptance rule and the model perturbations. The temperature decay rate for MMC

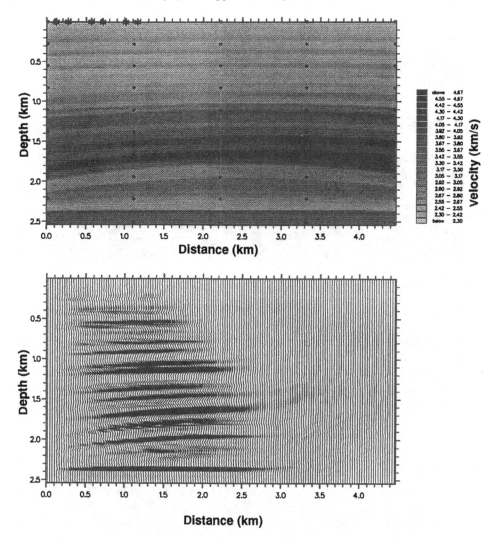

Figure 7.31 (a) 2D velocity model for evaluating MMC and RTC. The positions of the fifty spline nodes that will be used to parameterize this model are shown as black dots. The location of the three pairs of shot gathers used for MMC are shown as asterisks along the surface. The shot gathers to be modeled by RTC are the first of each pair. (b) Depth image formed by migrating the six shot gathers using the known velocity function. This is the best result that can be achieved for the data available.

was 0.98 and for RTC 0.88. Two trials per temperature were used. The decision to stop the algorithm was made when 40 iterations produced no significant change in the error. For MMC, this occurred after 230 iterations, which corresponds to 2,760 prestack depth migrations. For RTC, 334 iterations were used. RTC required a total of 668 poststack depth migrations and 2,004 forward model calculations.

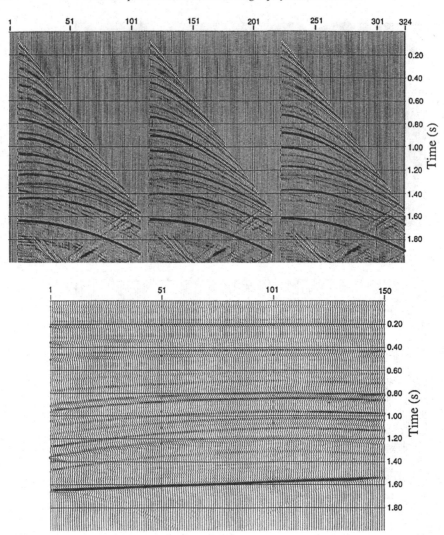

Figure 7.32 *(Above)* First shot gather of each shot pair after AGC and an external mute to eliminate wide-angle reflections and refractions. *(Below)* Zero-offset section corresponding to the velocity model of Figure 7.31. Depth migration of these data will be used to estimate the reflectivity for the synthetic shot gathers that RTC will compare with the observed data, which are shown in the upper part of this figure.

Figure 7.33 also shows the error associated with the best model estimated for each misfit criterion in terms of the other misfit criteria. For example, the best RTC model was used to calculate the migration misfit and is plotted as the cross in the MMC error plots (Figure 7.33, *left*). The best model obtained from MMC was used to compute the RTC error and is shown as an open circle in Figure 7.33 *(right)*.

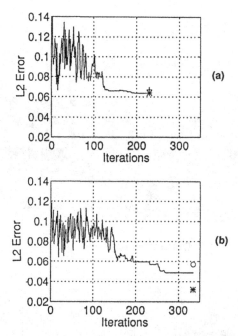

Figure 7.33 Plots of the error for (a) MMC and (b) RTC. Note that RTC still appears to be improving. In (b), the circle indicates the RTC value for the best model achieved by MMC, and in (a), the cross indicates the MMC value for the best model achieved by RTC. Asterisks indicate the best possible MMC and RTC that can be achieved when the known velocity function is used.

The best possible result for each measure that can be obtained using the known velocity function is also shown by an asterisk in both plots. This error is of interest because it is the minimum error possible. Since we are representing the true velocity model by splines, it is not likely that this error can ever be achieved. But the MMC inversion has reached approximately the same error as the migration misfit obtained using the true model and the best RTC model. Apparently, this measure of error finds all three models equally acceptable. Consequently, no further improvement in the MMC is likely without more data or possibly a different parameterization.

RTC, however, shows that the best result from MMC is not as good as the best RTC result, and the RTC result has not converged to the error obtained when using the known velocity function (see circle and asterisk, lower chart, Figure 7.33). It is obvious that RTC is a more sensitive measure of misfit than MMC, but the convergence of RTC is slower.

Figures 7.34 and 7.35 display the 2D velocity estimates for the best models derived from MMC and RTC, respectively. The images generated from summing the migrated shot gathers are shown below each velocity model. The MMC result

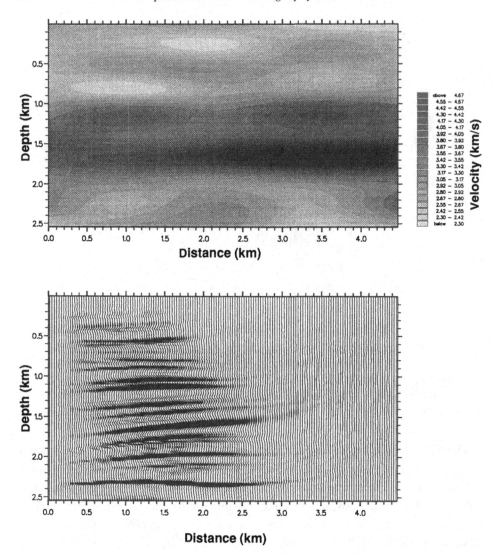

Figure 7.34 (a) Final velocity function obtained for MMC. The high-velocity zone has not been resolved very well (compare with Figure 7.31a). (b) Depth image formed by migrating the six shot gathers using the best MMC velocity function. The shallow part of the section is imaged correctly, but the reflections from the deepest part of the section have an incorrect dip (compare with Figure 7.31b).

(Figure 7.34) has not resolved the high-velocity zone as well as RTC (Figure 7.35). The seismic image for RTC compares very favorably with the image obtained using the known velocity model (Figure 7.31b), whereas the MMC image compares favorably only for the shallow part of the section. Staging over depth may improve the MMC result. Switching from MMC to RTC may prove to be an optimal

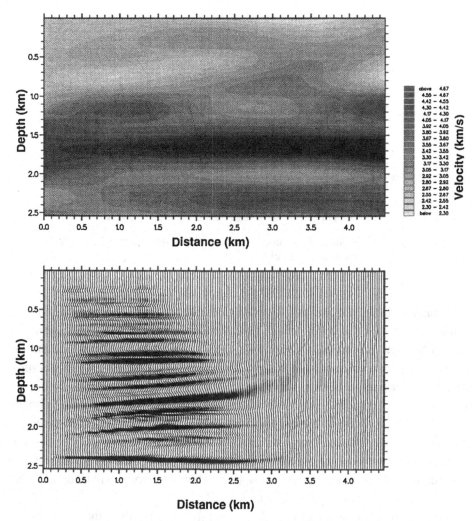

Figure 7.35 (a) Final velocity function obtained for RTC. The high-velocity zone has been resolved better than in Figure 7.34a. (b) Depth image formed by migrating the six shot gathers using the best RTC velocity function. The entire section is imaged correctly (compare with Figures 7.31b and 7.34b).

strategy that saves computational effort by using a lower-resolution objective function in the early evaluations and then a more critical measure of misfit after the MMC solution has converged.

7.2.3 Multiple and simultaneous VFSA for imaging

For the seismic velocity-analysis problem, it is difficult to use just the total misfit, either MMC or RTC, to represent the entire image. Since we know that the velocity

Local migration misfit

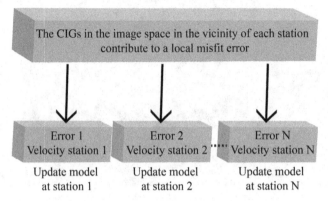

Figure 7.36 Subdivision of the model space: in this scheme, error corresponding to each station is updated separately.

is spatially variable, it makes sense to define regions of misfit based on subimages of the total. We did this by introducing velocity-analysis stations. The stations are defined by their x, y spatial location (which is included in the inversion) and the velocity–depth function at each station. These stations are then used to interpolate the sparser trial velocities onto the velocity grid used for imaging employing, for example, kriging. The error (MMC) is then defined using the common image gathers (CIGs) from these stations. The data used for each station's error estimate can overlap with the other station's data, and the station's position can be constrained to always have at least a minimum separation. Figure 7.36 shows the idea of subdividing the imaging space into stations where the model will be defined and around which the error will be estimated.

In this approach, one velocity model is used for imaging all of the seismic data. This model is defined from the velocities drawn at each station using that station's error and VFSA. Once the new models are drawn, they are interpolated to define the trial model for imaging. VFSA is in effect being run in parallel on each station's data, and the models being drawn independently are combined to form the new trial. The total error can also be found and used to judge the overall quality. We found this approach to speed up convergence of the velocity-analysis problem.

Figure 7.37 shows a 2s imaging example using this approach; see Stoffa *et al.* (2010) for details. Here we used an imaging algorithm that allowed us to trade horizontal resolution for speed. In the upper part of the figure, a low-spatial-resolution optimization was used to define an initial model. The CIGs are shown on the left, and the final image is on the right. This velocity was then used as the starting model for a more conventional resolution imaging, and the result is shown at

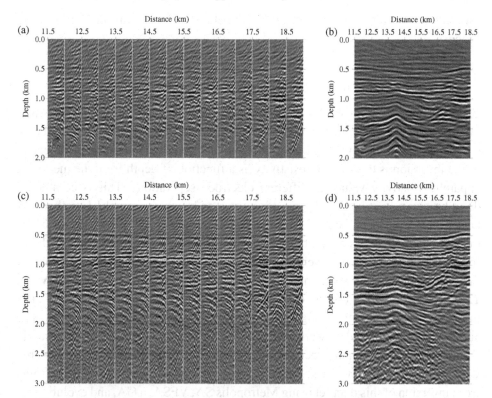

Figure 7.37 Multiple and simultaneous VFSA seismic imaging example. The CIGs are shown on the left, and the final image is on the right. The upper plot shows a low-spatial-resolution result used to obtain an initial velocity-depth model. This model was then used as a guide for more conventional resolution imaging, and the result is shown at the bottom. The CIGs are now better defined, and the final image (*right*) has more detail, including better fault definition.

the bottom of the figure. The CIGs are now better defined, and the final image (*right*) has more detail, including better fault definition. The velocity-analysis stations were located every kilometer along the line (11.5, 12.5, etc.) and used a 1.0-km aperture for the error estimates. Figure 7.37 shows the intermediate CIGs as well as the velocity station CIGs to show that the intermediate positions are well imaged by the distributed error VFSA algorithm.

This example shows the flexibility of the VFSA algorithm. We can independently solve smaller VFSA problems and combine their local models into a global model that is then evaluated. In the example here, any local imaged data will be sensitive to a large part of the global model being evaluated. The overarching considerations are the spatial variability of the subsurface image, the parameterization employed, and the number, frequency, and overlap of the stations. This approach can be applied to many problems where a local error estimate is suitable and can be used to contribute to the overall model-building process.

7.3 Inversion of resistivity sounding data for 1D earth models

The resistivity sounding method is commonly used in shallow subsurface investigations to map changes in formation resistivity as a function of depth. Measured data are the apparent resistivity values at different electrode separations. The objective of the inversion is to estimate resistivity as a function of depth from the measured apparent resistivity values at different electrode separations. This is commonly done by curve matching using master curves or by more formal LI methods. A heat bath SA has been applied to this problem by Sen *et al.* (1993). We observed that the heat bath SA is computationally very intensive for this problem and therefore describe results from inversion using other global optimization algorithms of the field Schlumberger sounding data (Bhattacharya and Patra 1968, table XVI, p. 127). The data, which consist of 19 measurements of apparent resistivity values in the electrode separation range of 2 m to 400 m, are shown in Figure 7.38. The authors noted that the data were collected with a Schlumberger electrode array in an area where sedimentary rocks with alternating beds of sand and clay, covered with a thin layer of alluvium, are expected to exist. In the following we show results from inversion of this data set using Metropolis SA, VFSA, a GA, and evolutionary programming (EP) employing the following mean square error function:

$$E\left(\mathbf{m}\right) = \frac{1}{ND} \sum_{i=1}^{ND} \left(\rho_o^i - \rho_s^i\right)^2,$$

$$\tag{7.11}$$

where ρ_o and ρ_s are the observed and synthetic apparent resistivity values, ND is the number of observation points, and $\mathbf{m}$ is the model vector consisting of resistivity and thickness of a set of layers. When the apparent resistivity values vary in a wide range, it is more appropriate to take the difference in the logarithm (base 10) of the synthetic and observed data for use in the error function. It is also straightforward to use any other type of error function. The heat bath SA inversion results will not be shown here, but they can be found in Sen *et al.* (1993).

7.3.1 Exact parameterization

It is quite evident from the plot of the data in Figure 7.38 that the data can be modeled with a three-layer earth model. Thus our model will consist of five parameters – three resistivity values and two thicknesses.

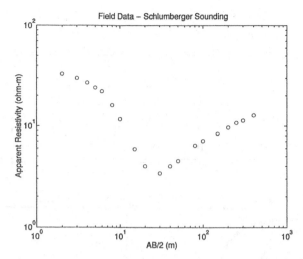

Figure 7.38 Schlumberger field sounding data to be used to illustrate different non-LI schemes. Note that the field data appear typical for a three-layer response.

Following Sen and Stoffa (1991), we call this parameterization an *exact parameterization*. We next define the search limits and search increments of the resistivity and thickness of different layers, as given in Table 7.1. Note that all three methods – Metropolis SA, VFSA, and EP – search continuously between the minimum and maximum values of each model parameter. The GA search is discrete, and therefore, the search increment shown in Table 7.1 is used only by the GA. We also note here that the search limits used are wider than those used by Sen *et al.* (1993) in the heat bath inversion of the same data set.

The Metropolis SA used a starting temperature of 0.1 and the cooling schedule shown by x in Figure 7.39a with 50 moves/temperature for 500 iterations, resulting in 25,000 forward modeling calculations. The VFSA used a starting temperature of 0.1 and a cooling schedule shown by o in Figure 7.39a with 10 moves/temperature for 100 iterations, resulting in 1,000 forward modeling calculations. The GA used a starting temperature of 1,000 with a temperature decay factor of 0.95 with 50 models for 77 generations (when the population became homogeneous) with a $p_x = 0.6$, $p_m = 0.01$, and $p_u = 0.9$. Thus the GA required 3,850 forward calculations to find an optimal answer. The EP used 50 models for 400 generations, resulting in 10,025 forward modeling calculations. Figure 7.39b shows the error-versus-iteration curves for both Metropolis SA and VFSA. The Metropolis SA attained a minimum error of 0.2903, and the VFSA attained an error of 0.2615. Figure 7.39c, d shows the minimum and average error at each generation of the GA and EP. Note that the GA plot (Figure 7.39c) shows an occasional increase in the average error

Table 7.1 *Search limits and search intervals used in the non-linear inversion*

Layer no.	ρ_{min} (Ω-m)	ρ_{max} (Ω-m)	$\Delta\rho$ (Ω-m)	h_{min} (m)	h_{max} (m)	$\Delta\rho$ (m)
1	10.0	100.0	1.0	1.0	20.0	1.0
2	1.0	20.0	0.1	1.0	20.0	1.0
3	1.0	20.0	1.0	—	—	—

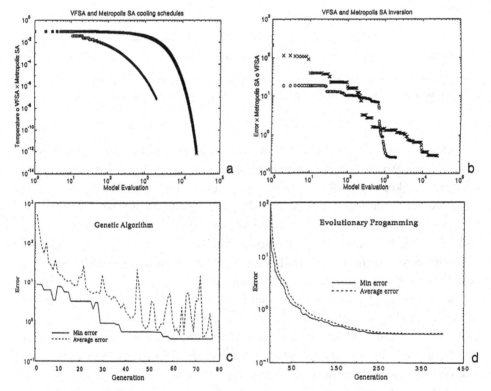

Figure 7.39 Inversion of resistivity sounding data by different non-LI schemes assuming a three-layer earth model (exact parameterization). (a) The cooling schedules used in VFSA (*open circles*) and Metropolis SA (*x*). Both algorithms started at the same temperature, but VFSA used a much faster cooling schedule. (b) Error-versus-iteration curves for VFSA (*open circles*) and Metropolis SA (*x*); VFSA finds a model with a very low error after only 1,000 model evaluations, whereas Metropolis SA took 10,000 model evaluations. (c) The dashed curve is the average error, and the solid line is the minimum error at each generation of the GA run. (d) The dashed curve is the average error, and the solid line is the minimum error at each generation of an EP run.

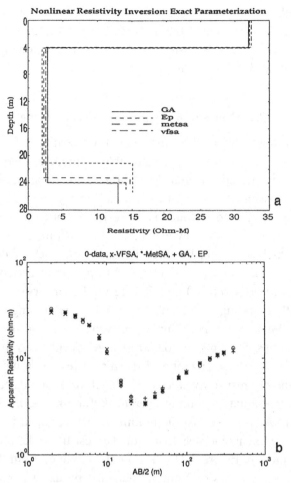

Figure 7.40 (a) The best-fit models obtained by VFSA, Metropolis SA, GA, and EP. The models are very similar except that there are some differences in the thickness and resistivity of the second layer. Since the final errors are very similar, this indicates the degree of uncertainty we can have in the thickness and resistivity of the second layer. (b) Comparison between field data and synthetic data predicted by the best-fit models obtained from the four inversion results shown in (a). They all are in good agreement.

(e.g., generation 45). This is so because one of the models in the generation had a very high error value, causing the average error of all the models to become high. Note also that since we are using a temperature-dependent selection probability, it prevents convergence of the population to the model that has a high error. The minimum error obtained by the GA was 0.358, whereas that found by EP was 0.343. The best model obtained by the four methods is shown in Figure 7.40(a). We notice that they are very similar. Comparison between observed and synthetic data from the best-fit models obtained by the four methods reveal that they are in very good agreement (Figure 7.40(b)). Thus all four methods were successful in modeling the data quite well. However, VFSA was found to be computationally the most efficient method, followed closely by the GA.

7.3.2 Overparameterization with smoothing

In the inversions shown in the preceding section, we restricted the number of layers to three. In general, it may not be possible to predetermine the number of layers. Therefore, we describe a more general model parameterization scheme in which a model is assumed to consist of several thin layers of constant thickness such that the only variable to be found is the resistivity of each layer. For the field data shown in Figure 7.38, we now use a 22-layer earth model with 10 layers of 1.0 m constant thickness each, followed by six layers each of thickness of 2.0 m, followed by six layers each of thickness 5.0 m. The resistivity of each layer was allowed to vary between 1.0 and 50.0 Ω-m. The inversion was carried out using VFSA with a starting temperature of 1.0 and the cooling schedule shown in Figure 7.41a. Fifty moves per temperature were used, and the algorithm was run for 600 iterations, resulting in 30,000 forward calculations. It is well known that this type of overparameterization of the model results in unrealistic models with highly oscillatory resistivity values. We therefore smoothed each model once using a simple three-point smoothing operator before the model was used in the evaluation by VFSA. We note that the minimum error attained by the VFSA was 0.356. The data and synthetic resistivity for the best-fit model are in reasonable agreement, as shown in Figure 7.41b. The model obtained from this inversion is compared with the best-fit VFSA model obtained by exact parameterization in Figure 7.41c. We observe that the first layer boundary can be easily identified, and the two models show a decrease in resistivity beneath the first layer, followed by an increase in resistivity with depth.

7.4 Inversion of resistivity profiling data for 2D earth models

We now illustrate the inversion of dipole–dipole resistivity profiling data for 2D resistivity variations in the subsurface. The simplest form of 2D resistivity variation is to assume some fixed geometric shapes of causative bodies such as circles, semicircles, vertical dikes, etc. for which synthetic resistivity data can be generated by analytic solutions. The inversion for such structures is rather simple because the number of model parameters is very small. For example, the only two model parameters for a vertical dike are the resistivity and thickness of the dike. Thus, for such a problem, the error function can be evaluated at each point

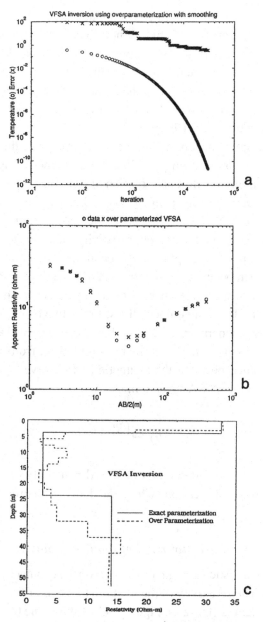

Figure 7.41 Inversion of field resistivity sounding data using an overparameterization of the model and smoothing: (a) the error **x** versus iteration and temperature (*open circles*) versus iteration are shown; (b) comparison between field data (*open circles*) and synthetic data predicted by the best-fit VFSA model; (c) comparison between the best-fit VFSA model obtained by exact parameterization (*solid line*) and overparameterization with smoothing (*broken line*). Note that even with overparameterization, the principal features of the model can be recognized.

in a 2D grid. Chunduru *et al.* (1995) showed that even for such simple structures, the resistivity inversion problem is highly non-linear and compared several of the non-linear inverse methods described in previous chapters and found VFSA to be the most computationally efficient. Here we will allow for a more general variation of resistivity in the subsurface such that the model cannot be approximated by simple geometric structures and use VFSA in the inversion of dipole–dipole resistivity profiling data (Chunduru *et al.* 1994).

Since here we want to accommodate general structures, the resistivity values should be allowed to vary at each point in a 2D grid, and the synthetic data must be computed using a finite-difference solution of Poisson's equation in two dimensions (Mufti 1976; Dey and Morrison 1979) or some variant of it. However, in our inversion, we represent the models by 2D cubic splines. This reduces the model space significantly and results in smooth resistivity models. The model parameters are the resistivity values at the spline node locations. The number of spline nodes and their locations can be either predetermined or solved for as a part of the inversion. The resistivity values at each grid then can be computed by interpolating between the spline nodes using cubic splines. In the two examples that follow, we have used a starting temperature of 100 with a decay rate of 0.98 and five moves per temperature. The search limits for the resistivity were from 1 to 150 Ω-m. The mean square difference between the synthetic and observed data is taken as the objective function in the VFSA inversion:

$$E = \frac{1}{ND} \sum_{i=1}^{ND} \left(R_s^i - R_o^i \right)^2, \tag{7.12}$$

where $\mathbf{R}_s$ and $\mathbf{R}_o$ are the synthetic and observed resistivity data vectors, respectively, ND is the number of locations where the data are recorded.

7.4.1 Inversion of synthetic data

First, we consider a realistic geologic model of a vertical contact covered by an overburden of 50 Ω-m. The resistivity of the medium to the left of the vertical contact is 10 Ω-m and to the right is 100 Ω-m. Figure 7.42(a) shows the true model. To test the validity of the spline node parameterization and its applicability, the inversion was applied to the data obtained from the vertical contact model. Even though the data were generated using the block model in Figure 7.42(a), for inversion purposes, the model was parameterized by cubic splines with six nodes in the x direction and four nodes in the z direction. The best model obtained after VFSA inversion is given in Figure 7.42(b). It can be seen that the best model obtained by VFSA has recovered the basic structure of the model along with the true resistivity values. The errors as a function of iteration are shown in Figure 7.42(c). In the initial iterations, the error is quite high. As the number of iterations increase, the error starts decreasing. Notice an occasional increase in error. This is due to the stochastic nature of the algorithm, where occasional increases in the error are acceptable. The pseudosection of the true data and the data obtained from the inversion are shown in Figure 7.42(d), (e). The pseudosection obtained from taking a difference between data and best-fit synthetic at each observation point is shown in Figure 7.42(f), which mostly shows very small values, indicating that there is close agreement between the original and the data predicted from the inversion result.

7.4.2 Inversion of field data

Next, we applied the VFSA inversion to a resistivity data set collected over a partially known zone of disseminated sulfide mineralization near Safford, Arizona. For the real-data inversion, we chose ten spline nodes in the x direction and five spline nodes in the z direction, respectively. The best models obtained from an LI and VFSA inversion (Narayan and Dusseault 1992) are shown in Figure 7.43a, b. The scale denotes the range of logarithmic resistivity values. Models shown in Figure 7.43a, b match closely with minor differences. The difference between models shown in Figure 7.43a, b are due to the difference in model parameterization. The best model proposed by Narayan and Dusseault (1992) (Figure 7.43b) is obtained by LI with blocky structures, where the shape of the subsurface structure was constrained. VFSA (Figure 7.43a) does not assume any prior information about the shape of the body but employs 2D splines, where resistivity values of the subsurface are allowed to vary from 1 to 90 Ω-m. Interestingly, the VFSA model appears like a smoothed version of the model proposed by Narayan and Dusseault (1992). The convergence of the error is shown in Figure 7.43c. In Figure 7.43d, the true data (+), the data obtained from VFSA inversion (*), and the data from the linear inversion of Narayan and Dusseault (1992) (o) are plotted. Overall, there is a good match between the real data and the data obtained from our inversion result. The minor differences in the results may be due to noise or limitations due to the sampling of the subsurface by fixed-shaped cells. For many data points, VFSA was able to obtain a better fit than the model of Narayan and Dusseault (1992).

7.5 Inversion of magnetotelluric sounding data for 1D earth models

The magnetotelluric (MT) method, first introduced by Cagniard (1953), uses surface measurements of natural electromagnetic fields to determine the resistivity (or conductivity distribution) of the earth. The data consist of apparent resistivity and phase measured at a wide range of frequencies (or periods). Most MT inversion methods aim at finding a resistivity (isotropic) as a function of depth only, although

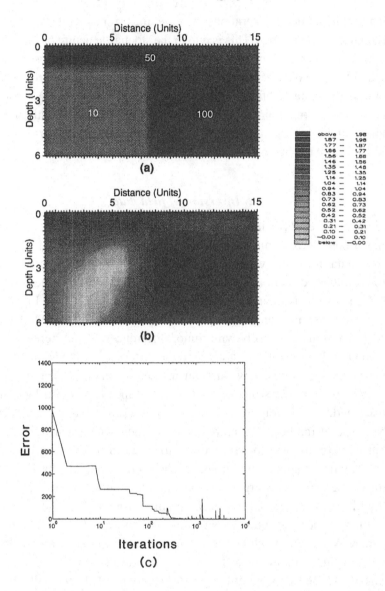

(a)

(b)

(c)

Figure 7.42 Inversion of synthetic dipole–dipole resistivity profiling data using VFSA; cubic splines were used to parameterize the model space: (a) the vertical contact model used in generating noise-free synthetic data for testing the inversion algorithm; (b) the best-fit model obtained from VFSA inversion using 2D splines to characterize the model, which closely resembles the true model – note that the models are plotted using a $\log_{10}$ resistivity scale; (c) error-versus-iteration plot for the VFSA inversion; (d) the data used in the inversion; (e) the synthetic data predicted by the best-fit VFSA model shown in (b) – note that they are in good agreement, as can be seen by inspecting the difference plot shown in (f).

APPARENT RESISTIVITY PSEUDO SECTION

```
       1----- 2----- 3----- 4----- 5----- 6----- 7----- 8----- 9-----10-----11-----12-----13-----14-----15-----

N=2        49.31  48.95  48.88  48.78  48.56  48.18  48.17  48.72  48.81  48.64  48.50  48.44

N=3           48.28  47.80  47.54  47.14  46.62  47.00  49.11  51.68  52.18  51.75  51.40

N=4              43.55  42.91  42.40  41.99  42.77  45.73  50.81  55.77  56.84  56.22

N=5                 36.88  36.16  35.99  37.09  40.21  45.68  53.37  60.61  62.44

N=6                    30.15  30.12  31.44  34.35  39.21  46.46  56.24  65.59

N=7                       25.29  26.66  29.25  33.25  39.14  47.64  59.13

N=8                          23.06  25.27  28.47  33.01  39.53  48.97

N=9                             22.43  24.93  28.36  33.17  40.14

N=10                               22.49  25.04  28.55  33.55

N=11                                  22.82  25.35  28.91          (d)

N=12                                     23.25  25.77
```

```
       1----- 2----- 3----- 4----- 5----- 6----- 7----- 8----- 9-----10-----11-----12-----13-----14-----15-----

N=2        49.44  49.42  49.26  48.79  48.52  48.54  48.77  48.71  48.16  47.60  48.13  50.82

N=3           46.58  48.00  47.19  45.32  44.81  46.54  49.45  51.47  51.75  51.32  51.96

N=4              43.51  43.48  41.57  40.34  41.97  46.28  51.39  54.91  56.37  57.08

N=5                 37.90  36.90  36.00  37.00  40.77  46.74  53.35  58.73  62.42

N=6                    31.24  31.25  32.43  35.20  40.08  47.11  55.62  63.94

N=7                       26.12  27.87  30.46  33.95  39.44  47.92  59.31

N=8                          23.31  26.08  29.08  32.88  39.34  50.10

N=9                             21.93  24.85  27.96  32.41  40.51

N=10                               21.03  23.89  27.43  33.08

N=11                                  20.35  23.48  27.93          (e)

N=12                                     20.15  23.99
```

```
       1----- 2----- 3----- 4----- 5----- 6----- 7----- 8----- 9-----10-----11-----12-----13-----14-----15-----

N=2        0.13   0.47   0.38   0.01   0.04   0.36   0.60   0.01   0.65   1.04   0.37   2.37

N=3           1.71   0.21   0.36   1.82   1.81   0.45   0.34   0.22   0.42   0.43   0.57

N=4              0.04   0.57   0.83   1.65   0.80   0.55   0.58   0.86   0.48   0.86

N=5                 1.02   0.74   0.01   0.09   0.56   1.06   0.01   1.87   0.02

N=6                    1.09   1.13   0.99   0.85   0.87   0.65   0.61   1.66

N=7                       0.83   1.22   1.21   0.70   0.29   0.28   0.17

N=8                          0.24   0.81   0.62   0.12   0.19   1.13

N=9                             0.50   0.07   0.40   0.76   0.37

N=10                               1.46   1.15   1.12   0.46

N=11                                  2.47   1.87   0.98          (f)

N=12                                     3.10   1.78
```

Figure 7.42 (*cont.*)

there has been some work on the inversion of MT data using 2D or 3D earth models. Although 1D earth models are simplistic, they still provide important information on earth properties. A comprehensive account of inversion of MT data for 1D earth models can be found in the monograph by Whitall and Oldenburg (1992). In this section we will report on the results from a Metropolis SA inversion of synthetic

MT data from a simple structure based on the work reported in Bhattacharya *et al.* (1992).

The forward problem for generating synthetic MT observations is well understood. For 1D earth models, the resistivity is assumed to be a continuous function of depth, or the earth can be modeled by several layers of varying thickness, each with a constant resistivity value. The former approach is described in Oldenburg (1979), and a description of the latter can be found in Porstendorfer (1975). Here we show an example of 1D MT inversion of synthetic data for a simple three-layer earth model. This model is the same as the one used by Oldenburg (1979) as an example in an iterative LI method. The noise-free synthetic data (shown by solid circles in Figure 7.44(b), (c)) in the period range 0.01 to 100 seconds were generated for the three-layered earth model shown in Figure 7.44(a). Here we are

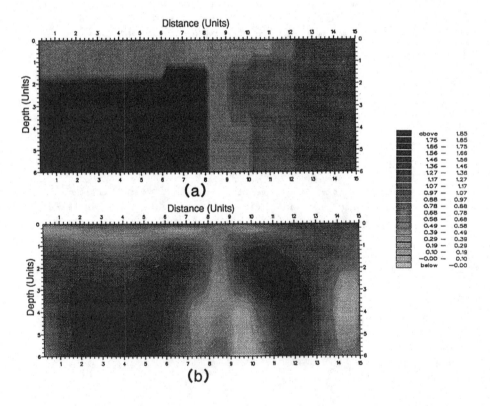

(a)

(b)

Figure 7.43 VFSA inversion of field dipole–dipole resistivity profiling data: (a) The best-fit model obtained from an LI scheme. (From Narayan and Dusseault 1992). (b) The best-fit model obtained from VFSA inversion. (c) Plot of error versus iteration from the VFSA inversion. (d) comparison between field data (*open circles*), data predicted by the best-fit VFSA model (**x**), and data predicted by the Narayan and Dusseault model (+).

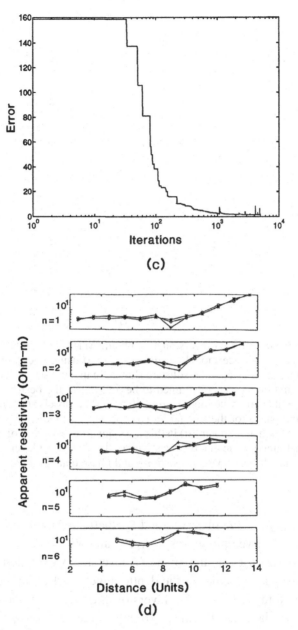

Figure 7.43 (*cont.*)

interested only in constructing a model that can explain the observation. For inversion using Metropolis SA, the following error function was minimized:

$$E = \frac{1}{ND} \sum_{j=1}^{ND} \left(\left(\rho_{obs}^i - \rho_{syn}^i \right) \middle/ \rho_{obs}^i \right)^2 + \sum_{j=1}^{ND} \left(\left(\Phi_{obs}^i - \Phi_{syn}^i \right) \middle/ \Phi_{obs}^i \right)^2, \qquad (7.13)$$

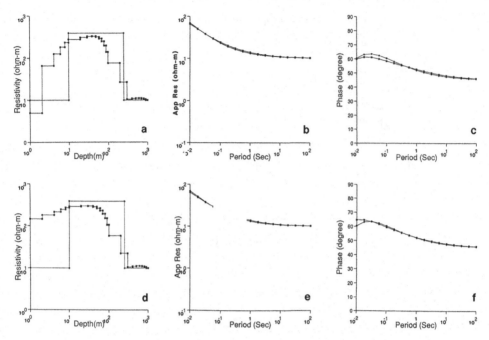

Figure 7.44 Metropolis SA inversion of synthetic MT sounding data for a simple two-layer-over-half-space model (*solid circles* in a) using overparameterization and smoothing. Two inversion results are shown. In the first set (a, b, c), the resistivity of the first layer was constrained to be 5 Ω-m: (a) the best-fit model (*solid squares*) resembles the true model (*solid circles*) very closely; (b, c) the apparent resistivity and phase for the true (*solid circles*) and the predicted (*solid squares*) models are in good agreement; (d) in this inversion, the resistivity of the first layer was not constrained. Note that the resistivity of the shallowest layer is very poorly resolved, although the apparent resistivity and phase data are matched very well (d, e).

where ρ_{obs} and ρ_{syn} are the observed and synthetic apparent resistivity and Φ_{obs} and Φ_{syn} are the observed phase, respectively, and ND (= 17 for this example) is the number of observation points. In the SA inversion, we divided the depth intervals linearly using a log scale for the depth range 1 to 1,000 m; i.e., we divided the depth range 0 to 10 m into five layers of thickness 2 m, the depth range 10 to 100 m into nine layers of thickness 10 m, and the depth range 100 to 1,000 m into nine layers of thickness 100 m. This resulted in 23 layers. In the SA inversion, we kept the thickness of the layers fixed and allowed the resistivity values to be searched in the range 5 to 1,000 Ω-m. This overparameterization of the model results in models (not shown) with oscillatory resistivity functions. This is due to the inherent non-uniqueness. To obtain geologically meaningful models, we attempted regularization by smoothing the models using a three-point operator before each model was evaluated. This simple scheme resulted in excellent results. For SA, an initial temperature of 1,000 was used, and the algorithm was run for 1,500 iterations, and 100 moves were made at constant temperature in each iteration. Results from two such SA inversions are presented in Figure 7.44. This figure shows the results from one such SA inversion in which the resistivity of the first layer was fixed at 5 Ω-m. The observed and synthetic data are in very good agreement, and the best-fit SA model also matches the main features of the true model. However, when the resistivities of all the layers were allowed to vary, the results are quite different. For example, the data are unable to resolve the thin surface layer with low resistivity (Figure 7.44a), although the apparent resistivity and phase data are matched quite well (Figure 7.44b, c).

7.6 Stochastic reservoir modeling

Recently, stochastic reservoir modeling that involves generating 2D or 3D images of reservoir rock properties using data from different sources such as core measurements, well logs, seismic and geologic information, well test data, etc. has become quite popular (e.g., Hewett 1986; Farmer 1988; Haldorsen and Damsleth 1990; Journel and Alabert 1990). Farmer (1988) gives a detailed description of the problem and methods for the solution. In reservoir description, we want values of an attribute such as porosity or permeability at each grid point in a 2D or 3D reservoir model. In practice, the problem is rather difficult because we are faced with the problem of inferring a large volume of data from a limited number of observations. This is why stochastic models for reservoirs are commonly used. In general, a large number of realizations or descriptions of the reservoir models will honor the available data. In practice, a few realizations are retained for analysis through flow simulation to study production schemes. Global optimization methods such as SA and GAs are appropriate for this problem because the problem of stochastic reservoir modeling can be cast as an optimization problem (e.g., Deutsch and Journel 1991; Sen *et al.* 1992), and an optimal configuration of the reservoir properties (i.e., porosity, permeability, etc.) can be obtained. One additional advantage of these algorithms is that they are not restricted to Gaussian random fields, and they can be used to accomplish geologic realism by combining data from different sources. In the following we describe an application of heat bath SA, Metropolis SA, and a GA to stochastic reservoir modeling based on Sen *et al.* (1992).

In this application, we attempt to obtain a permeability distribution in a reservoir in 2D (modeled by grids in x and z) such that the permeability is a spatially random variable with a known correlation structure defined by variograms $\gamma(h)$ that depend only on the separation distance h such that

$$2\gamma(h) = E\left\{\left[z(x) - z(x+h)\right]^2\right\},$$

$$(7.14)$$

where $z(x)$ is a spatially random variable (e.g., permeability at the grid blocks), and E is the expectation operator. Thus our problem reduces to that of creating a permeability field with a specified correlation structure such that the difference between the actual variogram and the variogram computed from the generated permeability field is minimum. We can use SA to minimize the following objective function:

$$e(\mathbf{m}) = \sum_i w_i \left[\gamma_c (h_i) - \gamma_\alpha (h_i) \right]^2,$$

(7.15)

where i varies over each of NX and NZ grid points along the x and z directions, respectively, $\mathbf{m}$ is the model vector of length $nx \cdot nz$ containing permeability values at the grid blocks, $\gamma_c(h)$ and $\gamma_a(h)$ are the computed and actual variograms, respectively, and w_i are a set of weighting factors.

Variograms can be calculated in as many directions as necessary. In our example, we consider a 2D reservoir model and try to minimize both the horizontal and vertical variograms. In our GA application, we define the following fitness function:

$$f(\mathbf{m}) = \frac{errtol}{\max\left[errtol, e(\mathbf{m}) \right]},$$

(7.16)

where $errtol$ is a prespecified error tolerance.

Using SA and GAs, we are not required to assume any functional form for the true variogram, so they can be used directly. However, it may be necessary to smooth the raw variograms, particularly when the data sampling is sparse. It is easy to incorporate conditioning data using SA and GAs simply by substituting the known permeability values at specific grid points and not allowing them to vary during the optimization. In our example (taken from Sen *et al.* 1992), we condition the data at the two end faces by assigning the known permeability values. This approach may produce discontinuities around the data, but this can be avoided by using a two-part objective function, as described by Deutsch and Journel (1991).

Here we show an example of the application of SA and GAs to the Antolini sandstone, an Eolian outcrop from northern Arizona. A rectangular slab of the sandstone of dimension $38 \times 13 \times 4$ cm was characterized by miniperameter measurements on each square centimeter. A contour plot of the measured permeability values for one face of the slab is shown in Figure 7.45(a), where we observe a low-permeability streak at the top and a high-permeability streak at the bottom. We also observe two additional high-permeability streaks in the middle of the rock slab. The permeability values range between 10 and 1,480 md with an average of 477 md. The horizontal and vertical variograms computed from these data are shown in Figure 7.45(b). We use the horizontal and vertical variograms of log-permeabilities in our simulations using SA and a GA because the experimental variograms are generally better behaved in the log-permeability domain than in the raw permeability domain.

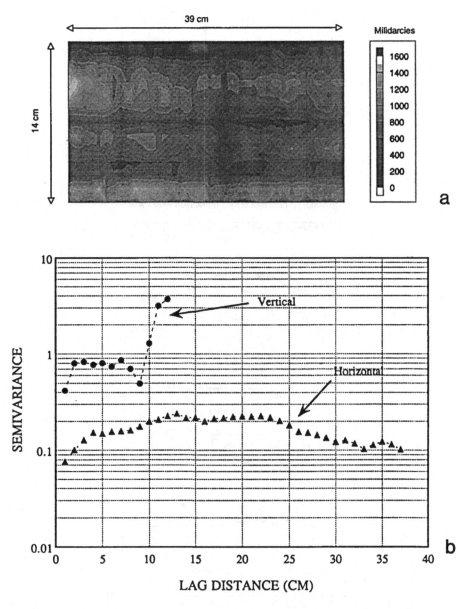

Figure 7.45 Stochastic reservoir modeling using SA and GA. (a) Contour plot of measured permeability values on one face of a slab of Antolini sandstone. The permeability values range between 10 and 1,480 md with an average of 477 md. (b) The horizontal and vertical variograms computed from the data shown in (a). SA and GA are used to find the permeability distribution that fits the measured variograms. (Reproduced from Sen *et al.* 1992; SPE 24754, presented at the 1992 SPE Annual Meeting, Washington, DC.)

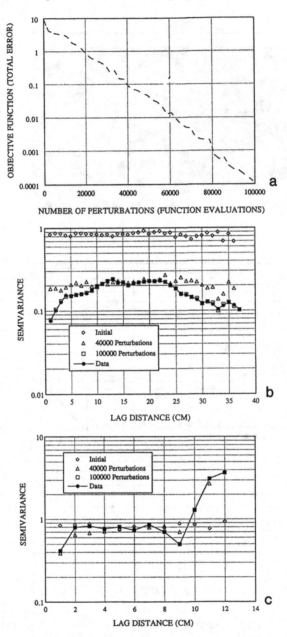

Figure 7.46 Simulation results using Metropolis SA. (a) Variation of error as a function of iteration. (b, c) The measured and reconstructed variograms at different stages of the simulation. (Reproduced from Sen *et al.* 1992 ; SPE 24754, presented at the 1992 SPE Annual Meeting, Washington, DC.)

Figure 7.46(a) shows the plot of error as a function of iteration for Metropolis SA using a cooling schedule of the type $T_n = 0.1(0.8)^n$, where n is the iteration number. One-hundred moves were allowed in each iteration of the Metropolis SA. The true and reconstructed vertical and horizontal variograms are shown at different stages of the optimization in Figure 7.46(b), (c). Clearly, the Metropolis SA reproduces the experimental variograms very closely.

The results from heat bath SA are shown in Figure 7.47(a), (b). The heat bath run was also restricted to 100,000 objective function evaluations, and a cooling schedule of the type $T_n = 0.1(0.8)^n$ was used in the simulation. The heat bath SA converged to the experimental variogram quite closely.

For the GA simulation, we used a population size of 200, and the GA converged after 1,000 generations (a total of 200,000 objective function evaluations) using a moderate value of crossover probability ($p_x = 0.6$), a small value of mutation probability ($p_m = 0.01$), and a high value of update probability ($p_u = 0.9$). The results from the GA simulation are shown in Figure 7.47(c). Although the agreement between the theoretical and experimental variograms appears good, we notice that the GA-generated horizontal variogram shows a slightly increased nugget effect.

Figure 7.48 shows the realizations of the permeability fields obtained using the Metropolis SA, heat bath SA, and GA. Comparison of these stochastic permeability fields with the actual data (Figure 7.45(a)) reveals that the major features of the outcrop sample are reproduced very well by all three methods. For example, the low-permeability layer at the top is present in each of the realizations, as well as in the data. Similarly, the high-permeability streaks are also reproduced in the proper locations.

Thus the Metropolis SA, the heat bath SA, and the GA were found to be quite useful in our simple example of reservoir modeling. Metropolis and heat bath SA required nearly the same amount of computation time, whereas the GA required twice as many objective function evaluations as those required by the other two methods. However, both the heat bath SA and the GA have significant potential for parallelization.

7.7 Seismic deconvolution by mean field annealing (MFA)
and Hopfield network

Reflection seismograms are usually explained using a model in which the source function is convolved with the impulse response of the earth to generate the observed seismograms. Seismic deconvolution is the process by which the effect of the source is removed in an attempt to recover the impulse response of the earth. This process helps to improve the temporal resolution of the seismic data. Commonly used statistical deconvolution methods impose certain assumptions

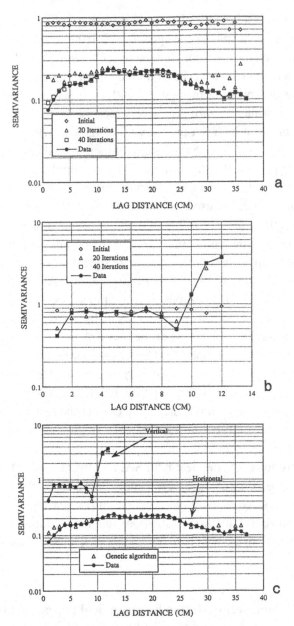

Figure 7.47 Simulation results using heat bath SA and GA. (a, b) The measured and reconstructed variograms at different stages of the simulation using heat bath SA. (c) The measured and reconstructed variograms at the final generation of the GA. (Reproduced from Sen *et al.* 1992; SPE 24754, presented at the 1992 SPE Annual Meeting, Washington, DC.)

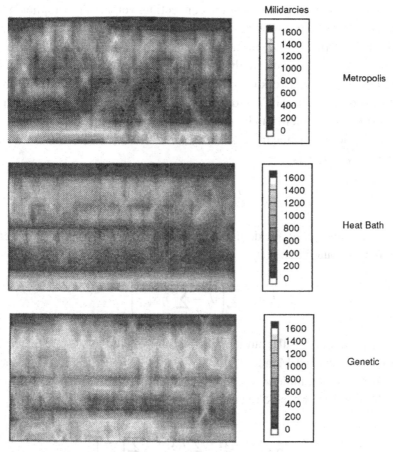

Figure 7.48 Sample realizations of the permeability distributions using Metropolis SA, heat bath SA, and a GA. Comparison with the measured data (Figure 7.43(a)) reveals that the major features of the outcrop sample are reproduced very well by all three methods. (Reproduced from Sen *et al.* 1992; SPE 24754, presented at the 1992 SPE Annual Meeting, Washington, DC.)

on the underlying earth model and the characteristics of the seismic source (e.g., that it is minimum phase) when there is no prior information available about the source wavelet. Other methods, such as minimum entropy deconvolution (Wiggins 1978; Ooe and Ulrych 1979) optimize some form of a non-linear function. Here, based on Calderon and Sen (1995), we describe a procedure that uses the least squares error criterion of Wiener predictive deconvolution in a Hopfield network (described in Section 4.6) and then employs MFA to escape from local solutions. This algorithm relaxes the minimum-phase assumption of the commonly used statistical deconvolution techniques.

We described in Section 4.6 how the connection weights of a Hopfield network are related to the forward problem through definition of the error function. In the deconvolution problem, the neurons can be used to represent the amplitudes of the reflectivity and source wavelet at different time samples (Wang and Mendel 1992). The task of the neural network is to find an optimal configuration of the neurons for which the error is minimum. We follow Wang and Mendel (1992) to describe how the deconvolution problem can be mapped to a Hopfield network.

Consider a single seismic trace given by the convolution of a source wavelet $s(t)$ with a reflectivity $r(t)$ as

$$z(t) = \sum_i r_i s(t - t_i),$$
(7.17)

where t_i is the time delay. We define the misfit or error between the observed trace $d(t)$ and the computed trace $z(t)$ as

$$E = \frac{1}{2} \sum_k \left(d_k - \sum_i r_i s_{k-i} \right)^2,$$
(7.18)

for $i, k = 1$ to N, and N is the number of time samples in the seismic trace.

Following Wang and Mendel (1992), r_i can be replaced by a 0–1 sequence x_i multiplied by an amplitude a_i, i.e., $r_i = a_i x_i$. Suppose that we set a_i equal to a constant α. Then the error can be rewritten as

$$E = \frac{1}{2} \sum_{ij} w_{ij} x_i x_j - \sum_i I_i x_i + \frac{1}{2} \sum_i \left(\frac{d_i}{\alpha} \right)^2,$$
(7.19)

where

$$w_{ij} = - \sum_k s_{k-i} s_{k-j},$$
(7.20)

and

$$I_i = \sum_k s_{k-i} \frac{d_k}{\alpha} - \frac{1}{2} \sum_k s_{k-i}^2,$$
(7.21)

for $i, j, k = 1, \ldots, N$, and $w_{ij} = 0$. Comparing the energy function (7.20) with that of a Hopfield network (Eq. 4.47), i.e.,

$$E = -\frac{1}{2} \sum_{ij} w_{ij} x_i x_j - \sum_i x_i \theta_i,$$

we notice that Eqs. (7.20) and (7.21) are the connection weights and threshold, respectively.

Assuming a known source, this network can be used to detect a set of amplitudes $\{S_\alpha\}$ of magnitude proportional to α by iteratively solving for x using the following (Eq. 4.46) local update rule:

$$x_i(n+1) = H\left[\sum_{ij} w_{ij} x_j(n) + I_i\right],$$

where n is the iteration number.

To compute the amplitudes a_i with $i \in \{S_\alpha\}$, it is necessary to code a_i in such a way that it can be represented by bipolar units. With that purpose, a_i is approximated with the following binary representation:

$$a_i = \sum_{j=1}^{M} \frac{1}{j-1} y_{ij} - 1, \quad i \in \{s_\alpha\}, \tag{7.22}$$

where y_{ij} is either 0 or 1, and M gives the discretization or resolution of the variable a_i in the range $(-1, 1)$. Using this representation for the amplitudes in the error equation, we arrive at the following equation:

$$E = \frac{1}{2}\sum_k \left[d_k - \sum_{i\in\{s_\alpha\}} s_{k-i}\left(\sum_j \frac{1}{2^{j-1}} y_{ij} - 1\right)\right]^2,$$

which is simplified to the following form:

$$E = -\frac{1}{2}\sum_{ij}\sum_{nm} w'_{ij,nm} y_{ij} y_{nm} - \sum_{ij} I'_{nm} y_{ij} + \text{constant}, \tag{7.23}$$

with

$$w'_{ij,nm} = -\sum_k \frac{1}{2^{j+m-2}} s_{k-i} s_{k-n},$$

and

$$I'_{ij} = \sum_k \left[\frac{1}{2^{j-1}}\left(d_k + \sum_{l\in\{S_\alpha\}} s_{k-l}\right) s_{k-i} - \frac{1}{2(4^{j-1})} s_{k-i}^2\right],$$

where $k = 1, \ldots, N$; $i, n \in \{S_\alpha\}$; $j, m = 1, \ldots, M$; and $w_{ij,nm} = 0$. This network can now be used to obtain the amplitudes of the detected reflections.

At this stage the description of the problem in terms of a Hopfield network is complete. Wang and Mendel (1992) used a local update rule iteratively to compute the reflectivity amplitudes and their respective delay times. The algorithm starts with a large value for α and computes the time delays and then the corresponding amplitudes. Next, the seismic trace is updated by removing the computed reflectivity, α is decreased by a small quantity, and the process is repeated until even the small amplitudes have been removed from the seismic trace. This process alternately sweeps positive and negative amplitudes of the seismogram until it finishes at the signal level that is considered noise in the data.

7.7.1 Synthetic example

Here we initially demonstrate the application of both the local update rule and MFA to the deconvolution of a synthetic trace, assuming that the source function is known. Figure 7.49a shows a non-minimum-phase wavelet that is convolved with a spike series (Figure 7.49b) to generate a seismogram (Figure 7.49c). The objective here is to estimate the spike series (Figure 7.49b) from the seismogram using the wavelet. (Note that this is a simple problem, and several methods can be used for this purpose. We chose this simple example only to illustrate the algorithm. Figure 7.49d–f shows the spike series locations and the amplitudes at different stages of MFA deconvolution. Recall that we run different passes of the algorithm with gradually decreasing values of amplitude first to locate the spike positions and then to derive an amplitude at those spike locations. First, a value for α that represents the maximum amplitude of the signal is selected ($\alpha = 1$). The located spikes using this value of α are shown by plus sign (+) in Figure 7.49d. The predicted amplitudes for these spikes are marked by 0 in Figure 7.49e. The final result from MFA is shown in Figure 7.49f along with the true reflectivity series; they are exact. The reflectivity series obtained at the final iteration of a local update rule–based deconvolution is shown by asterisks in Figure 7.49g. Clearly, we obtained superior results using MFA.

The process to compute the source wavelet is essentially the same as the one used to compute the amplitudes for a single α, considering that the reflectivity r_i is known. The only difference is that for each time sample there is an amplitude associated with it.

7.7.2 Real-data example

We now show an example of deconvolution of real data in which we estimate both the source function and the reflectivity series using MFA. The iterative algorithm is described in Figure 7.50. First, an initial guess of the wavelet is used, and estimates of the reflectivity are made using MFA until a minimum error or convergence is reached. Next, we hold the reflectivities fixed and use MFA to update the source function. The alternation between reflectivity and source function is repeated until convergence is attained.

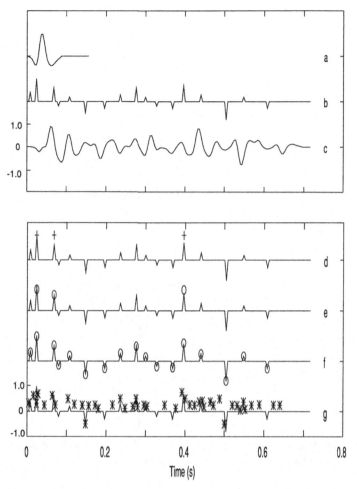

Figure 7.49 Deconvolution of a synthetic seismic trace by MFA: (a) source function; (b) reflectivity series; (c) synthetic seismic trace obtained by convolution of the spike series with the source function; (d) result from the first step of MFA deconvolution (the + signs show the locations of the spikes, with the highest amplitude, detected by MFA; (e) amplitudes of spikes detected by MFA marked with 0; (f) final MFA inversion result (the reflectivity amplitudes predicted by MFA are marked with 0s); (g) result from a local update rule–based deconvolution. The predicted reflectivity amplitudes are marked with asterisks.

One CMP gather (CMP 1602) from the Carolina Trough (Wood 1993) is shown in Figure 7.51. Note that the data have been move-out-corrected for the water layer. Figure 7.52a shows initial wavelets at each offset used in the MFA deconvolution. These minimum-phase wavelets were obtained from a window of 0.12 second of the autocorrelation of each trace and are reasonable initial estimates for the source functions. Three iterations of MFA were performed to arrive at the final solutions.

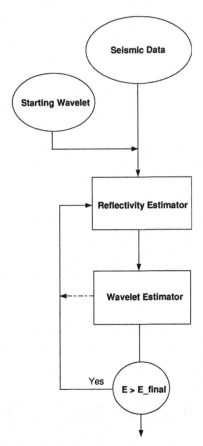

Figure 7.50 Processing steps for deconvolution by MFA and Hopfield networks. A starting wavelet is necessary to initiate the process.

The final wavelets are displayed in Figure 7.52b. Figure 7.52c, d presents expanded plots of three of the initial and final wavelets, respectively. We notice that the final wavelets have more late-arriving energy than the initial wavelets. The reflectivity series estimated by the MFA is shown in Figure 7.53a. The predicted seismograms obtained by convolution of the recovered spike series shown in Figure 7.53a and the source functions (Figure 7.52b) are shown in Figure 7.53b. The difference between the predicted (Figure 7.53b) and the observed seismograms (Figure 7.51) are shown in Figure 7.53c. These residuals show the data not modeled with this convolution model and show incoherent energy with very small amplitude, probably representing noise in the data. Thus we find that MFA can be used successfully in the seismic deconvolution problem to estimate source wavelets and the reflectivity series.

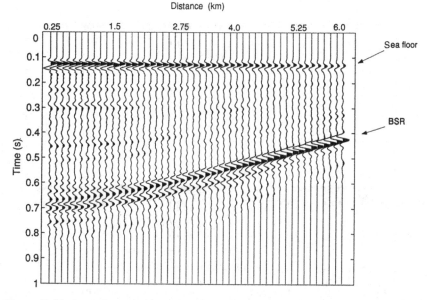

Figure 7.51 A marine seismic gather from the east coast of the United States (the Carolina Trough). The data have been move-out-corrected for the water layer.

7.8 Joint inversion

Non-linear optimization methods such as VFSA are readily employed when more than one data type needs to be included in the optimization problem. We have two geophysical examples that illustrate the method. The first is the combination of first-arrival travel-time picks from deep-water ocean bottom seismic data and simultaneously acquired marine gravity data. The second is the combination of seismic data with oil field production data. Our first example is real geophysical data, so the actual answer is unknown. Our second example is a controlled numerical experiment where we know the correct answer.

7.8.1 Joint travel time and gravity inversion

For the travel-time and gravity problem, we use a functional relation between the velocity of the rocks in the subsurface and their density. Roy *et al.* (2005) describe in detail the method we employed. The density $\rho(x, z)$ is related to the velocity $\alpha(x, z)$ through a polynomial with known coeficients a_j, where $j = 1, 5$. Thus it is totally dependent on the velocities that will come from the seismic data. In this way, the gravity data act as a constraint on the velocity inversion. We could have solved for the coefficients of the polynomial as well, but for simplicity, they remain as known model parameters:

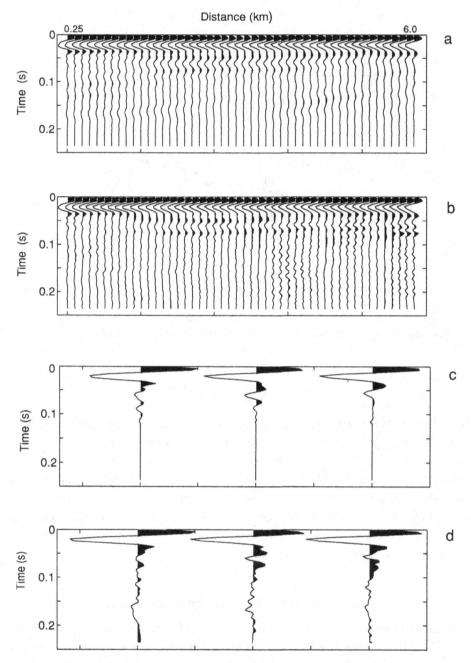

Figure 7.52 (a) Initial minimum-phase source wavelets used in the MFA deconvolution; (b) final source wavelets predicted by the deconvolution process; (c, d) the expanded plots of three of the initial and final wavelets, respectively.

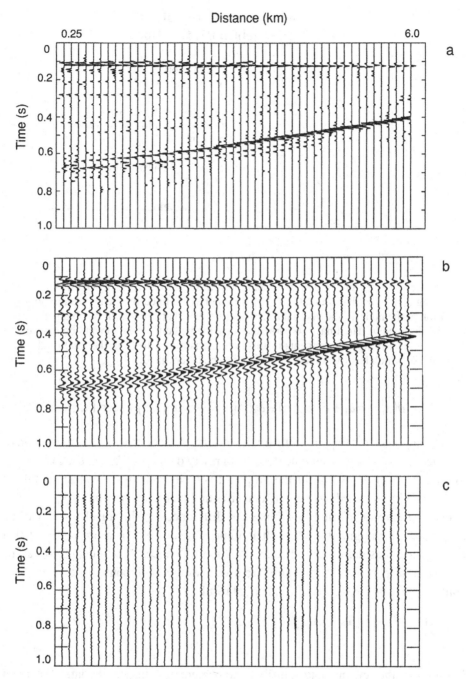

Figure 7.53 (a) Reflectivity series predicted by MFA; (b) predicted seismograms; (c) residual between observed and predicted data.

Model parameterization
Arc-tangent basis functions

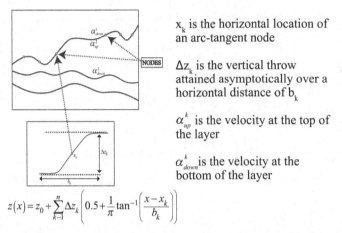

x_k is the horizontal location of an arc-tangent node

Δz_k is the vertical throw attained asymptotically over a horizontal distance of b_k

α_{up}^k is the velocity at the top of the layer

α_{down}^k is the velocity at the bottom of the layer

$$z(x) = z_0 + \sum_{k-1}^{n} \Delta z_k \left(0.5 + \frac{1}{\pi} \tan^{-1} \left(\frac{x - x_k}{b_k} \right) \right)$$

Figure 7.54 Parameterization scheme for a subsurface velocity-depth model using arctangent basis functions. The geometry is defined by the horizontal position of the node x_k, the vertical throw dz_k referenced at x_k that will be obtained by the distance b_k. The final two parameters are the velocity at the top and bottom of the layer. Multiple arctangents of this type are summed to form the final model, each of which has five model parameters to be determined.

$$p(x,z) = a_1 + a_2 \alpha(x,z) + a_3 \alpha^2(x,z) + a_4 \alpha^3(x,z) + a_5 \alpha^4(x,z). \qquad (7.24)$$

The input data for this problem are the observed travel times T_{obs} as a function of distance from the source and the observed gravity data g_{obs}. We will use ray tracing to predict travel times T_{syn} for the proposed velocity structures we test, and these trial velocities are used to compute densities and then gravity values g_{syn}.

Our measure of misfit E combines the differences in the travel times and the differences in the gravity values. It is important to use a measure of misfit that is normalized, as described earlier, so that the influence of each data type can be controlled as part of the procedure. We used a weight w to control the level of the contribution of each data type to the final result:

$$E\left[\alpha(\mathbf{x}), \rho(\mathbf{x}) \right] = \left\| \mathbf{T}_{obs} - \mathbf{T}_{syn} \right\|^T C_T^{-1} (\mathbf{T}_{obs} - \mathbf{T}_{syn}) + w \left\| \mathbf{g}_{obs} - \mathbf{g}_{syn} \right\|^T C_g^{-1} (\mathbf{g}_{obs} - \mathbf{g}_{syn}).$$

$$(7.25)$$

In this case, the seismic data contribute 80 percent and the gravity data 20 percent to the total error. Using VFSA, we can select any measure of misfit we believe appropriate as long as the maximum error is 1.0 and the minimum is 0.0. Here we used the normalized error energy measure.

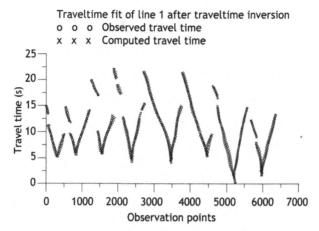

Figure 7.55 Predicted travel time T_{syn} and observed travel times T_{obs}. The predicted travel times are the result of the inversion using only the travel-time data and the arctangent paramaterization for the velocity model. The agreement is excellent.

To reduce the number of model parameters, we approximate the subsurface using a parameterization scheme that can generate complex models but which does so with a reasonable number of parameters. We chose arctangents as basis functions (see Figure 7.54). Each arctangent node has five model parameters that need to be determined. The first is the horizontal position of the node x_k. The second is the vertical throw dz_k referenced at x_k that will be attained by the distance b_k. These describe the geometry, and the final two parameters are the velocity at the top and bottom of the layer. Figure 7.54 shows in detail how these five parameters can be used for multiple nodes to define a reasonably complex velocity model.

Figure 7.55 shows the actual T_{obs} and predicted T_{syn} travel times when the optimization was performed without including gravity data. At this scale, they are essentially identical. But when we look at the predicted gravity data (Figure 7.56), we see that there are large discrepancies between the predicted gravity and observed gravity. To reconcile both observations, a joint inversion was performed with the seismic data contributing 80 percent to the error. Figure 7.57 shows the observed and predicted travel times. They do not agree as well as the travel-time-only inversion, but overall, the fit is still quite reasonable. Figure 7.58 shows the new predicted gravity compared with the observed gravity, and now there is much better overall agreement. A better fit may be obtained by relaxing the constraint on the coefficients of the polynomial that defines density from velocity. That is, these could be free parameters that are solved for in the inversion. Also, the number of arctangent nodes could be increased to improve resolution.

Gravity corresponding to velocity model

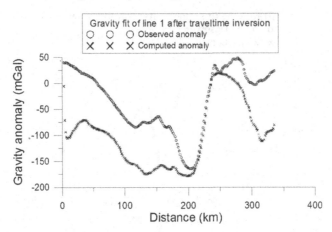

Figure 7.56 The velocity model estimated from just using the travel times (see Figure 7.59 upper) was then used to compute the gravity response. Here we compare the observed gravity g_{obs} and the predicted gravity g_{syn}. The predicted gravity is far from the observed gravity, indicating that the density model derived from this velocity model is incorrect.

Result: Joint inversion travel time data

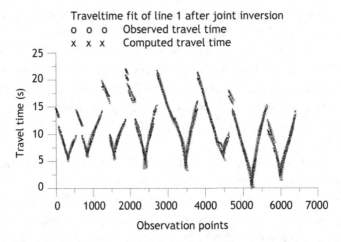

Figure 7.57 Travel-time comparison for the inversion done with both the travel times and the gravity data. It is clear that this result does not match the travel time as well as the optimization that used only the travel-time data (Figure 7.55). This indicates that the gravity data are influencing the result.

Figure 7.59 compares the crustal velocity models derived from just the travel-time inversion (*above*) and the joint travel time and gravity inversion (*below*). At first glance, they appear quite similar, which is not surprising. But when you look in detail, there are significant differences. The inclusion of the gravity data has generated a more rapid drop of layer 8 near OBS 13, and the shallow (–5 km) basin in the travel-time-only model (100 to 220 km along the line) has now been

Joint inversion: gravity fit

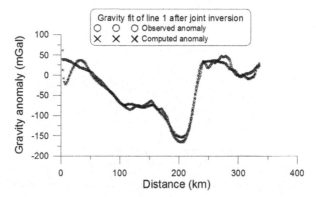

Figure 7.58 A comparison of the observed g_{obs} and predicted g_{syn} gravity data after the joint inversion of both travel-time and gravity data. The agreement is much better than that obtained from the travel-time data optimization that did not include gravity (Figure 7.56), as expected.

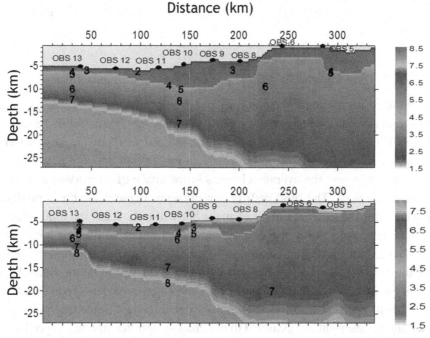

Figure 7.59 The upper plot is a display of the velocity model obtained for the inversion using just the travel-time data. The lower plot shows how the velocity model is changed when the optimization includes both the travel-time data and the gravity data. The OBS locations along the line are numbered to show the positions and the number and location of the seismic receivers along this profile. Color version available online at www.cambridge.org/sen_stoffa.

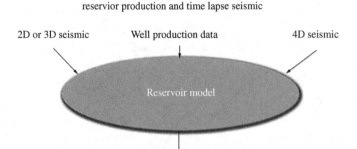

MAP coupled joint inversion using VFSA:
reservior production and time lapse seismic

2D or 3D seismic   Well production data   4D seismic

Reservoir model

Prediction of future performance
and production strategies optimization

Figure 7.60 Using repeated 3D seismic surveys acquired over a multiyear period along with well production data, we can try to predict reservoir properties, and these then become useful for guiding production strategies. The problem is how to combine these very different data types.

replaced by a smaller basin (180 to 220 km). The deepest structure – layer 7 (250 to 350 km along the line) – is also significantly shallower when the gravity data are included.

7.8.2 Time-lapse (4D) seismic and well production joint inversion

The combined inversion of seismic data and well production data has the potential to refine estimates of reservoir properties and thereby guide future production strategies (Figure 7.60). The problem is that these data types are very different in the way they respond to reservoir changes. Production data are acquired only at wells and represent the average changes in the space of a reservoir as they are detected at the well. The fluid distribution changes as the reservoir is produced. Well data will be sensitive to these changes but will only sense the total or aggregate change because these observations are at a limited number of locations in space. The seismic data are acquired in 3D. That is, data are acquired in the area above the reservoir zone and, after processing, are imaged to depth. Spatial resolution is on the order of tens of meters over the entire area surveyed. This acquisition is then repeated at, for example, one-year intervals. Since the rocks above and within the reservoir are assumed to not change significantly due to production, changes in this now 4D or time-lapse seismic response are most likely to be due to changes in the fluids as the reservoir is produced.

So the data available for this problem are imaged 3D seismic data acquired at multiple time lapses and the well production data, which is continuously measured, but at only a few spatial positions. We illustrate the potential of using an optimizing method such as VFSA to solve the following problem: Given ten years

Porosity inversion procedure honoring both static and dynamic data for one scale

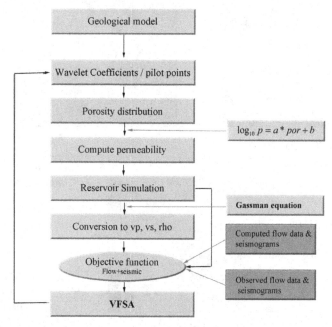

Figure 7.61 Flowchart of the 4D seismic and well production optimization problem to determine porosity. A sparse model parameterization of the porosity using either wavelets or pilot points is used to define a plane at the top of the reservoir. The reservoir simulations are run, and the results are used to predict well production and seismic data. Both these data sets are used to define a combined error, which is then used by VFSA to draw a new porosity model.

of production data and ten 3D seismic surveys imaged to depth, can we estimate the porosity of the reservoir rocks? The porosity does not change with production, but the saturation does. This results in a change in the seismic response over time. As in the preceding example, we need to define some rules that relate the physical properties. (In the preceding example, we linked density to velocity using a polynomial.) Here permeability must first be linked to porosity. Then, using a rock physics model, the reservoir properties (i.e., porosity and saturation) must be linked to the compressional-and shear-wave velocity and density to which seismic data are sensitive. The permeability porosity relation used (Figure 7.61) is well established, and some flexibility can be included by also solving for the two coefficients a and b. Here they are kept fixed. The rock physics model used is Gassman's equation, although other models could be employed, either in different inversion runs or optionally as part of one inversion to generate several possible seismic responses for evaluation.

Evaluate gridded model based on pilot points

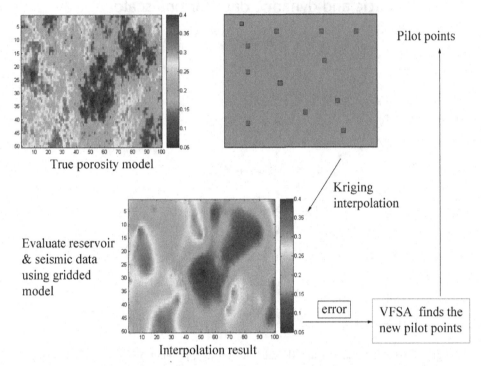

Figure 7.62 The model parameters we estimate in this example are pilot-point positions and porosity values. Fifty pilot points (three model parameters per pilot point) were used, and then kriging was used to interpolate these onto the computation grid that is used for the reservoir simulation and the seismic data. Color version available online at www.cambridge.org/sen_stoffa.

In this example (see Jin *et al.* 2009 for details), the seismic data will be just the imaged normal-incidence reflection coefficient in one plane at the top of the reservoir. We use VFSA to define a porosity model to be tested, define permeability from the porosity, and then using a reservoir simulator, produce the reservoir over a ten-year period. Each year reservoir properties are recorded, and we use the rock physics model to generate the seismic data that would be observed. We measure the misfit between observed and predicted seismic data, and observed and predicted well production data.

In this example we parameterized the porosity using pilot points. We start by using our known porosity and generate the synthetic "real" data against which we compare our results. We follow the sequence of operations described in Figure 7.61. For each iteration, we use VFSA to define new pilot points. In this case, there are three parameters per pilot point: the x and y positions and the porosity. Kriging is used to generate all the intermediate spatial values on the grid needed for

Inversion results

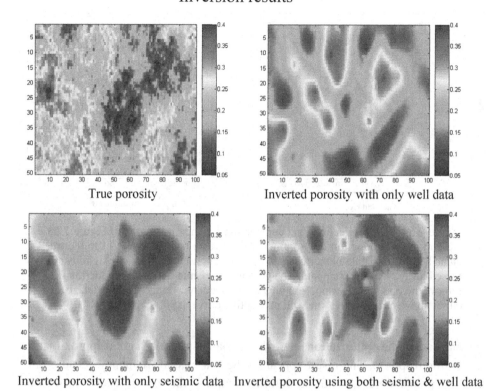

True porosity   Inverted porosity with only well data

Inverted porosity with only seismic data   Inverted porosity using both seismic & well data

Figure 7.63 Three porosity estimates from VFSA optimization. The true porosity is shown in the upper left. The upper right shows the result of using just the
well production data. The lower left shows the porosity estimate when only the
seismic data are used. The lower-right result was obtained using both the seismic
and well production data. Color version available online at www.cambridge.org/
sen_stoffa.

the reservoir simulation and the seismic data (Figure 7.62). The error is defined
by summing the squared differences in the observed and test seismic and well
production data. These are combined using a weight of 80 percent for the seismic
data in our best results.

Figures 7.63 and 7.64 show how these two very different data types affect the
optimization results. In the upper left of Figure 7.63 we show the true porosity.
In the upper right is the result when only the production data are used. It is clear
that the estimated spatial distribution of the porosity has little or no correlation
with the true porosity. Yet in Figure 7.64 this result agrees almost perfectly with
the true well production data. It should be clear that if the location of a new well
were needed, the well production data alone could not provide any useful information. In the lower left of Figure 7.63 we have the result of using just the 4D

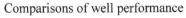

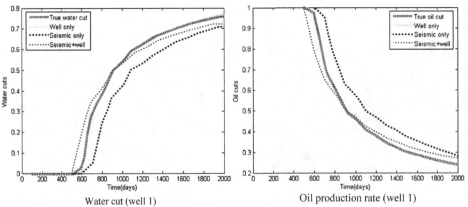

Figure 7.64 Well production data comparison for the three cases of Figure 7.63. Here the well production-only optimization gives a nearly perfect fit to the true data. The seismic data alone result in well production data that are significantly biased away from the true data. The combined seismic and well production data remove the bias in the seismic-only result and seem to average the true data, being either above or below the true well production curve at 900 days.

seismic data in the optimization. Here the spatial response is a good approximation to the true response. The result appears to be low spatial frequency because only 50 pilot points were used in the optimization, and these were constrained to be within equal-area boxes to keep them arealy distributed. Figure 7.64 demonstrates that the seismic data alone do a poor job of predicting the well production data, as we would expect. When we do the joint inversion with the errors weighted at 80 percent for the seismic differences, we get the result shown in Figure 7.63 (*lower right*) and Figure 7.64, where the predictions now seem to average the observed data.

There are many opportunities to expand the scope of this application. This very simple example shows the power of sampling methods such as VFSA to find reasonable answers to what would otherwise be very difficult problems. The two data types used here are very different and are sensitive to changes in different ways. Yet we can combine both data types to constrain our answers and get results that are very good approximations to the real situation.

8

Uncertainty estimation

In Chapter 2 we described several approaches to formulating geophysical inverse problems. It was pointed out that the geophysical inverse problem of estimating earth model parameters from observations of geophysical data often suffers from the fundamental limitation that several models may fit the observations very well. This phenomenon, which has been called *non-uniqueness* in the geophysical literature, may be caused by several factors, some of which were described in Section 2.3. The most well recognized of these is that the real earth properties vary continuously in all spatial directions (i.e., the model space is truly infinite-dimensional), and we are faced with the problem of constructing an earth model from a finite – albeit small – set of measurements. Thus the inverse problem is highly underdetermined and will result in many non-unique solutions (e.g., Menke 1984). Since in many cases the earth may be modeled with a discrete (small) set of layers based on independent information, this type of non-uniqueness can be greatly reduced. One other cause of non-uniqueness is related to the problem of identifiability or sensitivity of the model to the data. In seismic tomography problems, the regions of the earth with little or no ray coverage cannot be resolved from the data, and the slowness estimates for these regions will be ambiguous. Also, any noise present in the data or use of inexact theory to predict a model from the data may also cause a large degree of uncertainty in the results. Thus an interpreter is faced with several choices for the earth model, and often the number of choices can be reduced based on prior knowledge of the earth models.

Several attempts have been made to reduce uncertainty by imposing constraints such as regularization or solving for smooth models. Of course, what kind of smoothing is appropriate is highly debatable. In global tomography problems, use of spherical harmonics has been very popular (Dziewonski 1984). Both spherical harmonics and cubic splines are smoothing operators and may introduce unrealistic features in the model (Shalev 1993), if not used properly. The optimal fitting of splines or any other function to gridded models is itself an optimization problem. However, they have often been found to generate geologically meaning-ful solutions.

One other approach of describing the inverse problem is to use a statistical frame-work and to attempt to describe or characterize the non-uniqueness of the solution by describing the solution in terms of the probability density function (pdf) in the model space. This approach is valid for any model parameterization that we prefer to choose. In many situations, we may have prior information to restrict the models to a small set of parameters, but even then different model parameter values either may be altered independently or may depend on other parameters to explain the observed data. The statistical approach enables us to estimate uncertainty bounds on the resulting model and the correlation between different model parameters. Such an approach has also been criticized for the requirement of prior knowledge of data and theory errors. The advantages of the statistical approach are that it results in the marginal posterior probability density function (PPD) of the model given the observed data and several measures of uncertainty in the model space can be obtained for a given parameterization. Even though most statistical approaches make simplistic assumptions of Gaussian priors and uncorrelated data errors, the results obtained from such approaches are often physically meaningful.

The Bayesian formulation was described in detail in 2.6, and it was shown that the PPD $\sigma(\mathbf{m}|\mathbf{d}_{obs})$ of earth model $\mathbf{m}$, where $\mathbf{d}_{obs}$ are the measured data, describes the solution of the geophysical inverse problem when a Bayesian inference model is used. In many applications, the PPD is neither analytically tractable nor easily approximated, and simple analytic expressions for the mean and variance of the PPD are not available. Since complete description of the PPD is impossible in the highly multidimensional model space of many geophysical applications, several measures, such as the highest posterior density regions and the marginal PPD, and several orders of moments are often used to describe the solutions. Calculation of such quantities requires evaluation of multidimensional integrals. Here we will describe several different numerical integration schemes that can be used in the evaluation of such multidimensional integrals with application to geophysical inversion (Sen and Stoffa 1994).

The very basic problem of drawing samples from PPD is that there exists no simple analytic form of the PPD in general, and the PPD is highly multidimen-sional. A class of algorithms called *Markov chain Monte Carlo* (MCMC) addresses these issues. Recall that the original Metropolis algorithm proposed in the early 1950s is essentially a sampling algorithm in which a sample is first drawn from a trial (or proposal) distribution. This sample is then accepted or rejected using the Metropolis criterion. As shown in Chapter 5, after a large number of trials (infin-ite) at a constant temperature, the algorithm converges to a stationary distribution, which happens to be the PPD in our application. The proof makes use of the theory of Markov chains that satisfies the properties of ergodicity and aperiodicity. A similar proof was also discussed in Chapter 4 for the heat bath algorithm. In other words, Chapter 4 builds the fundamental concepts of MCMC. It has now been shown that for a large class of trial solutions, the algorithm converges to a stationary distribution (e.g., Gregory 2005, ch. 12). Several papers have been published in the geophysical literature on the application of several variants of MCMC to geophysical inversion (e.g., Mosegaard and Tarantola 1995; Sen and Stoffa 1996; Curtis and Lomax 2001; Malinverno 2002; Mosegaard and Sambridge 2002; Sambridge and Mosegaard 2002; Buland and More 2003; Malinverno and Briggs 2004; Malinverno and Leaney 2005). In the following sections we will describe these ideas, primarily based on Sen and Stoffa (1996).

8.1 Methods of numerical integration

In Section 2.5 we described the Bayesian formulation of the geophysical inverse problem in detail. Bayes' rule is a very convenient tool with which to update our current knowledge when new information becomes available. It describes the solution of an inverse problem by means of the PPD of model $\mathbf{m}$ given the observed data $\mathbf{d}_{obs}$, which is a product of a likelihood function and prior pdf $p(\mathbf{m})$ of model $\mathbf{m}$ (Eq. 2.53). The likelihood function is a function of data misfit. Recall that the PPD (Eq. 2.145) is given by the following equation:

$$\sigma(\mathbf{m}|\mathbf{d}_{obs}) \propto \exp[-E(\mathbf{m})]p(\mathbf{m}). \tag{8.1}$$

The pdf $p(\mathbf{m})$ is the probability of the model $\mathbf{m}$ independent of the data; i.e., it describes the information we have on the model without the knowledge of the data and is called the *prior* pdf. Similarly, the pdf $\sigma(\mathbf{m}|\mathbf{d}_{obs})$ is the description of the model $\mathbf{m}$ for the given data and is called the *posterior* pdf or the PPD when normalized.

The choice of the likelihood function depends on the distribution of the noise or error in the data (Box and Tiao 1973; Cary and Chapman 1988). Thus it requires prior knowledge of the error distribution. This is a very important issue because in many situations it is very difficult to obtain an estimate of noise statistics. The error can be due to measurement (e.g., instrument errors) or due to the use of inexact theory in prediction of the data (Tarantola 1987). Assuming Gaussian errors, the likelihood function takes the following form:

$$l(\mathbf{d}_{obs}|\mathbf{m}) \propto \exp[E(\mathbf{m})], \tag{8.2}$$

where $E(\mathbf{m})$ is the error function given by

$$E(\mathbf{m}) = \left(-\frac{1}{2}\left[\mathbf{d}_{obs} - g(\mathbf{m})\right]^T \mathbf{C}_D^{-1}\left[\mathbf{d}_{obs} - g(\mathbf{m})\right] \right), \tag{8.3}$$

where g is the forward modeling operator, and $\mathbf{C}_D$ is called the *data covariance matrix*. The data covariance matrix consists of two parts, namely, the experimental uncertainty or observational error $\mathbf{C}_d$ and the modelization uncertainty or error due to theory $\mathbf{C}_T$ (Tarantola 1987), and it is assumed that the errors due to theory and observation are independent.

By substituting a Gaussian pdf for the prior $p(\mathbf{m})$, we can derive an equation for the posterior pdf very easily. Note, however, that even under the assumption of Gaussian priors, the posterior pdf is non-Gaussian due to the presence of the term $g(\mathbf{m})$ in the likelihood function in Eq. (8.2). The expression for the PPD (Eq. 2.72) thus can be written as

$$\sigma(\mathbf{m}|\mathbf{d}) = \frac{\exp\left[-E(\mathbf{m})\right]p(\mathbf{m})}{\int d\mathbf{m}\,\exp\left[-E(\mathbf{m})\right]p(\mathbf{m})}, \tag{8.4}$$

where the domain of integration spans the entire model space.

Once the PPD has been identified, as given by Eq. (8.4), the answer to the inversion problem is given by the PPD. Even if the PPD were known, there is no way to display it in a multidimensional space. Therefore, several measures of dispersion and marginal density functions are often used to describe the answer. The marginal PPD of a particular model parameter, the posterior mean model, and the posterior model covariance matrix are given by Eqs. (2.147), (2.148), and (2.149), respectively. All these integrals can be written in the following general form (Eq. 2.150):

$$I = \int d\mathbf{m}\, f(\mathbf{m})\, \sigma(\mathbf{m}|\mathbf{d}_{obs}). \tag{8.5}$$

Thus the marginal PPD, posterior covariance, and mean model can be calculated by numerical evaluation of integrals of the type given in Eq. (8.5). We will review four different approaches to evaluating these integrals: grid search, Monte Carlo integration, Monte Carlo importance sampling, and an approximate method based on multiple maximum a posteriori (MAP) estimation.

8.1.1 Grid search or enumeration

This approach is the simplest and most straightforward. The entire region in the multidimensional model space in which the integrals need to be evaluated is discretized and divided into uniform grids. Then the error function, such as the one given by Eq. (8.3), is evaluated at each point in the model space. Thus the marginal posterior PPD, covariance, and mean model can be evaluated using numerical integration techniques such as the trapezoidal rule. The integration result depends on the grid spacing and the method of numerical integration employed (Press *et al.* 1989). For most geophysical inversion problems, the model space is usually very large, and the forward modeling is very computationally intensive. Since the grid-search technique requires evaluation of the error function (i.e., the forward calculation) at each point in the model-space grid, it is impractical for most geophysical applications.

8.1.2 Monte Carlo integration

Unlike the grid-search method of numerical integration, in which the integrand is evaluated at points in a uniform grid over a multidimensional model space, Monte Carlo integration schemes (Section 1.8) make use of pseudo-random number generators. The integrand is evaluated at points chosen uniformly at random. Although there are some rules (Rubinstein 1981), in practice, it is difficult to estimate how many function evaluations will be required for accurate estimation of the integral.

8.1.3 Importance sampling

The Monte Carlo method draws samples (random vectors) from a uniform distribution, many of which do not significantly contribute to the integral. The idea of importance sampling (Section 1.9) is to concentrate the selection of sample points from the regions that are the most important, i.e., which contribute the most to the integral. This requires that unlike the pure Monte Carlo method, the random vectors must be drawn from a non-uniform distribution. Such a non-uniform distribution cannot, however, be chosen arbitrarily. Hammersley and Handscomb (1964) and Rubinstein (1981) showed that for an unbiased estimation of the integral with minimum variance, we require that the samples of $\mathbf{m}$ be drawn from the pdf $\sigma(\mathbf{m}|\mathbf{d}_{obs})$. The estimation of the integral will be biased if the models are drawn from a distribution other than $\sigma(\mathbf{m}|\mathbf{d}_{obs})$. For many geophysical applications, the prior $p(\mathbf{m})$ is far from being sharply peaked, and the PPD is clearly dominated by the likelihood function. Thus, generating the models according to the prior alone is not enough, and importance sampling by drawing models according to the PPD is essential. This, however, is not a trivial task. Recall that the PPD $\sigma(\mathbf{m}|\mathbf{d}_{obs})$, as given in Eq. (8.4), contains an integral in the denominator that requires that the error function $E(\mathbf{m})$ be evaluated at each point in model space. Clearly, there is no need for importance sampling if we know the value of the error function at each point in model space. In Section 8.2 we discuss some ways of drawing models from $\sigma(\mathbf{m})$ without evaluating the error function at each point in the model space. Once the models are drawn accordingly, the frequency distribution of each model parameter represents marginal PPDs for each model parameter.

8.1.4 Multiple MAP estimation

Several MAP estimation algorithms can be developed based on global optimization methods such as simulated annealing (SA) and genetic algorithms (GAs). To reach near the global minimum of an error function, these methods sample different parts of the model space. Several independent runs of SA, GAs, or very fast simulated annealing (VFSA) can be made; the sampled models must be weighed by their corresponding un-normalized PPD (Eq. 8.4), and then integrals of the type given in Eq. (8.5) should be evaluated. The algorithms should be run repeatedly with different starting solutions until the estimates of marginal PPD and posterior and mean model do not change. This method assumes that the PPD is simple and well behaved (not necessarily Gaussian) and that by repeating several different runs, we are able to adequately sample the most significant portion of the PPD. In general, this approach will result in biased estimates of the integrals. However, the accuracy of the method will depend on the shape of the PPD, the optimization method used and the number of independent runs used in the optimization method.

8.2 Simulated annealing: the Gibbs sampler

Chapter 4 described SA methods in general and outlined several different algorithms. Chapter 7 showed several different geophysical applications of SA where the principal objective was to find a model that matches the observations the best. In this section we will show how SA can be used as an importance sampling tool.

As we saw in Chapter 4, there are two basic computer algorithms of SA: Metropolis SA (Metropolis *et al.* 1953; Kirkpatrick *et al.* 1983) and the heat bath algorithm (Rothman 1986). Although both the algorithms simulate the same phenomenon, they differ substantially in the details of implementation. For example, Metropolis SA is a two-step procedure in which a model is drawn at random and then a decision is made whether to accept or reject it. The generation-acceptance process is repeated several times at a constant temperature. Next, the temperature is lowered following a cooling schedule, and the process is repeated. The algorithm is stopped when the error does not change after a sufficient number of trials. On the other hand, heat bath SA is a one-step procedure. This algorithm computes the relative probability of acceptance for each possible move before any random choice is made; i.e., it produces weighted selections that are always accepted. In this approach, the model parameters are discretized to a desired accuracy. Each model parameter is visited sequentially for all possible values of one model parameter while keeping the values of all other model parameters fixed.

Thus the Metropolis and the heat bath SA algorithms differ substantially. It has been suggested that heat bath may be faster for models with a large number of parameters (Rothman 1986). One important difference between the two methods is that heat bath works with discrete values of each model parameter, whereas Metropolis SA does not. Therefore, for small increments in model parameter values, the heat bath algorithm may be slow.

Although the two algorithms differ, they have one thing in common. Each can be modeled as a finite Markov chain that is irreducible and aperiodic (see Chapter 4). Consequently, it can be shown that at a constant temperature, after a large number of iterations, they attain an equilibrium or a steady-state distribution that is independent of the starting model. The equilibrium distribution at a temperature T attained by both Metropolis and heat bath SA algorithms is given by the following Gibbs pdf:

$$p(\mathbf{m}) = \frac{\exp\left[-\dfrac{E(\mathbf{m})}{T}\right]}{\displaystyle\sum \exp\left[-\dfrac{E(\mathbf{m})}{T}\right]}, \tag{8.6}$$

where the sum is taken over all the models in the model space.

Thus we have achieved an important goal of the importance sampling discussed in Section 8.3. That is, the models must be sampled according to a distribution such that we do not need to evaluate the error function at each point in model space. The Markov chain analysis of SA guarantees that – in equilibrium – the models are sampled according to a Gibbs distribution. The proof, however, is asymptotic, meaning that we require an infinitely large number of iterations at a constant temperature. In practice, it is possible to attain such a distribution in a finitely large number of iterations, which is smaller than that required by a grid search. We will show this numerically for a geophysical problem in Section 8.5.

Since SA samples models from the Gibbs distribution, such a sampler is called a *Gibbs sampler* (GS). Strictly speaking, a Metropolis based sampling is called a Metropolis–Hastings MCMC approach, while sampling based on a heat bath algorithm is called a Gibbs sampler. A GS based primarily on the heat bath algorithm and some of its variants has been applied recently to several problems outside the field of geophysics (e.g., Gelfand and Smith 1990; Gelfand *et al.* 1992; Gilks and

Wild 1992) in computing posterior marginal distributions. In geophysics, such an approach has been discussed in Basu and Frazer (1990). Gibbs sampling based on the heat bath SA will involve sweeping through each model parameter several times at constant temperature. Similarly, a Gibbs sampling based on the Metropolis SA will involve repeating the model generation-acceptance procedure using the Metropolis rule a large number of times at a constant temperature. It may also be useful to make several independent GS runs using different starting models and random-number seeds to avoid dependence on the computer-generated random numbers. The two crucial parameters are the number of iterations for each independent run and the number of independent runs necessary to achieve convergence (Gelfand and Smith 1990). The first parameter is the number of iterations necessary to reach convergence of the error-versus-iteration curve. The second parameter can be determined based on how long it takes to obtain stable estimates of the marginal density functions, mean, and posterior covariances. In principle, the heat bath and Metropolis rule–based GSs are equivalent, but we prefer the latter over the former primarily because the heat bath–based GS works with discrete values for each model parameter and can be quite slow for a large number of model parameters where each model parameter is discretized very fine.

At this stage it is worthwhile to re-examine the PPD given by Eq. (8.1). A comparison with Eq. (8.6) reveals that the PPD is essentially the Gibbs pdf at temperature $T = 1$ for uniform priors. This means that to evaluate quantities such as the marginal PPD, mean, and covariance by Monte Carlo importance sampling, we need to sample the models using a GS at a constant temperature of 1. The frequency distribution of the model parameters directly give the marginal PPDs. We also note here that the original Metropolis *et al.* (1953) algorithm was actually invented as a Monte Carlo importance sampling technique (Hastings 1970) to study the equilibrium properties, especially ensemble averages, time evolution, etc., of very large systems of interacting components, such as molecules in a gas. Following Metropolis *et al.* (1953), we obtain samples $\mathbf{m}_1, \mathbf{m}_2, \ldots, \mathbf{m}_{NM}$ such that the mean can be approximated by the following ergodic average:

$$\langle \mathbf{m} \rangle \cong \frac{1}{NM} \sum_{j=1}^{NM} \mathbf{m}_i,$$
(8.7)

and the covariance will be given by

$$C_M \cong \frac{1}{NM} \sum_{j=1}^{NM} \left(\mathbf{m}_i - \langle \mathbf{m} \rangle \right) \left(\mathbf{m}_i - \langle \mathbf{m} \rangle \right)^T,$$
(8.8)

where NM is the number of models.

8.3 Genetic algorithm: the parallel Gibbs sampler

The concepts of the GA were described in detail in Chapter 5. Recall that a GA works with a population of models that are coded in some suitable form, and the algorithm seeks to improve the fitness (which is a measure of the goodness of fit between data and synthetics for the model) of the population generation (iteration) after generation. This is accomplished by the genetic processes of selection, cross-over, and mutation. A modified GA that incorporates some of the concepts of SA (Stoffa and Sen 1991) was described in Chapter 5.

We showed in Section 5.5 that most GAs are not based on rigorous mathematical models. We therefore examined the Stoffa and Sen (1991) algorithm by considering a simple but less rigorous model with the aim of comparing it with classical SA.

In our analysis, we follow a particular member of the population through different generations. Here a trial consists of the processes of selection, crossover, mutation, and update. A model is first generated by the processes of selection, crossover, and mutation. Then its fitness is compared with that of a model from the preceding generation. The new model is always accepted if the fitness of the new model is higher. When the fitness is lower, it is accepted using the well-known Metropolis acceptance criterion.

The model of the GA described in this manner (Eqs. 5.28 and 5.29) is very similar to that used in describing Metropolis SA, except that the generation probability in Metropolis SA is uniform, whereas GA used a biased selection. This causes problems when we attempt to describe the GA in terms of Markov chains. Although for a finite mutation rate the Markov chain may be considered ergodic (i.e., there is finite probability of reaching every state from every other state), the selection probability as given by Eq. (5.10) is controlled by the partition function in the denominator, which takes on different values for different generations. However, when the selection temperature T_s is very high, the models will be selected uniformly at random from the population. In such a situation, if the crossover rate is zero and the mutation rate is high, the Stoffa and Sen (1991) GA is the same as running several Metropolis SAs in parallel. However, if we consider a high T_s, moderate crossover ($p_x = 0.5$), and moderate mutation ($p_m = 0.5$), the crossover mixes models between different parallel SAs, and mutation allows for a local neighborhood search. Strictly speaking, such an approach will cause a breakdown of the symmetry property of the generation matrix, which is essential to prove convergence to a stationary distribution. However, due to the presence of a high mutation rate, we expect that when such an algorithm is run for a large number of generations at a constant acceptance temperature, it will attain a Gibbs distribution at the acceptance temperature. We will call this approach a *parallel Gibbs sampler* (PGS) and compare the results from PGS and GS for a geophysical example in the following section. Thus our "recipe" for a PGS is as follows: Run several repeat GAs with different initial populations with T_s very high and an acceptance temperature T_u of 1.0 with $p_x = 0.5$ and $p_m = 0.5$. Note that due to this high mutation rate, the population will never become homogeneous, but a single GA can be stopped when the variance of the fitness of the population becomes nearly a constant. The number of parallel GAs will be determined by examining when the estimates of the marginal PPD, mean, and covariance become stable. The frequency distribution of the model parameters sampled by these runs will directly estimate the marginal PPDs, and the posterior mean and covariance will be computed by the formula given in Eqs. (8.7) and (8.8).

8.4 Numerical examples

8.4.1 Inversion of noisy synthetic vertical electric sounding data

In this section we investigate the validity of different numerical integration algorithms using an example from vertical electric sounding (VES) data. VES data were chosen for this exercise because of the fact that the forward calculation is very rapid, and several methods can be tested quite rapidly. Details of VES inversion by non-linear method are described in Sen *et al.* (1993). Here we generated synthetic VES data for a three-layer earth model (Figure 8.1a). Ten percent Gaussian random noise was added to each synthetic data point. Thus the data shown in Figure 8.1b are a single realization of a random process. Note that for this example, the data are uncorrelated, and the prior data covariance matrix is known. In the inversion we will use exact theory. The only contribution to C_D data covariance matrix comes from the data error. An error function as given in Eq. (8.3) is used in all the inversion runs. The objective of this exercise is not just to obtain a best-fit model but to characterize the uncertainties by means of marginal PPDs, mean, and covariances, as given by Eqs. (8.7) and (8.8). These will be evaluated by several different methods, namely, a grid search, GS, PGS, and multiple MAP estimation, algorithms based on VFSA, multiple Metropolis SA, and multiple GAs. For all these, the search window and search increments used in the process are given in Table 8.1. Note that the discrete search was only used by grid search, GA, and PGS.

The calculation by grid search is straightforward. Forward calculations were done for each point in the discrete grid given in Table 8.1, and Eqs. (8.7) and (8.8) were evaluated using the trapezoidal rule. This resulted in 669,600 forward calculations. The GS (in this case a single Metropolis SA at $T = 1.0$) was run until convergence was achieved in the estimation of marginal PPDs. For PGS, we used 50 models and 10 parallel GA runs at a constant acceptance temperature of 1.0. Each

Numerical examples

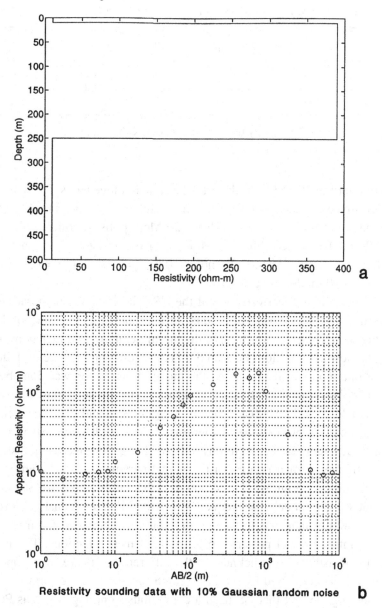

Figure 8.1 (a) A three-layer, 1D resistivity model used to generate synthetic Schlumberger vertical electric sounding data to test different sampling-based uncertainty-estimation algorithms. (b) Resistivity sounding data for the model shown in Figure 8.1a with 10 percent Gaussian random noise added to each data sample.

Table 8.1 *Search window and search increments for the resistivity and thickness*
of the three layers used in the inversion by several algorithms

ρ_{min}(Ω-m)	ρ_{max}(Ω-m)	$\Delta\rho$(Ω-m)	h_{min}(m)	h_{max}(m)	Δh(m)
5.0	15.0	1.0	1.0	20.0	1.0
300.0	450.0	10.0	150.0	400.0	10.0
1.0	20.0	1.0	—	—	—

Note: The discrete search was only used by grid search, GA, and PGS.

GA was run until the variance of the error became homogeneous. We will see that 10 parallel runs were adequate. On the other hand, 20 multiple VFSA runs (with 12,000 models) were evaluated. Similarly, 20 Metropolis SA runs (with cooling) with 20,000 models were evaluated. Multiple GAs consisted of 10 parallel runs each with a population of 50 models and $p_x = 0.6$, $p_m = 0.01$, and $p_u = 0.9$ (constant) resulted in 22,000 model evaluations.

First, we describe the performance of the GS. The error-versus-iteration curve is shown in Figure 8.2. Note that a constant acceptance temperature of 1.0 was used in this case. The error was high in the initial iterations and then converged to sampling the low-error regions beyond 5,000 iterations (in this case each iteration consists of one model evaluation, and the model perturbation was done by changing one model parameter at a time). Histograms of the model parameters sampled at iterations 10,000 and 50,000 are shown as Figure 8.3a and b, respectively. Note that beyond iteration 50,000, the histograms no longer changed. Also, repeating the GS with different starting solutions did not change the frequency distribution either. The frequency distributions of the model parameters at different stages of the PGS are shown in Figure 8.4a (24,000 models) and Figure 8.4b (80,000 models). Beyond 80,000 models, the frequency distribution did not change further.

Next, we simply map the frequency distributions shown in Figure 8.3b and Figure 8.4b into marginal posterior probability density functions. The marginal PPDs for the resistivity and thickness of the different layers are superimposed on the marginal PPD computed by the grid-search method and are shown in Figure 8.5. Considering the fact that GS used a continuous sampling, whereas PGS and grid search used discrete sampling between the maximum and minimum values of model parameters given in Table 8.1, the match between the three estimators is good. Note that the GS results were binned at the same discrete interval as that used by PGS and grid search.

In Figure 8.5, we also compare the marginal PPDs for the five model parameters evaluated by multiple Metropolis SA, multiple GAs, and multiple VFSA,

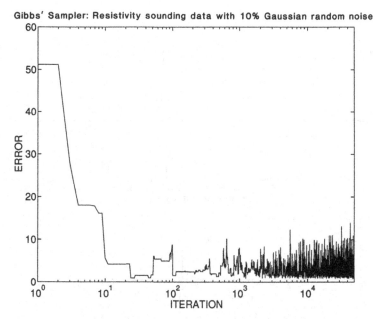

Figure 8.2 Error-versus-iteration curve for a Gibbs sampler at a constant temperature of 1.0 in the inversion of VES data shown in Figure 8.1. The error was initially very high, but beyond iteration 5,000, it sampled low-error regions of the model space. Here each iteration corresponds to one model evaluation.

respectively, with those obtained from the GS, PGS, and grid search. We notice that – overall – the width of the distribution obtained by the multiple MAP algorithms is slightly narrower than the width obtained by the Gibbs sampler. This means that the posterior variance of these parameters will be slightly underestimated by the multiple MAP algorithms. This is expected because MAP algorithms are biased toward sampling densely around the maximum of the PPD.

We also note that the plots of the marginal PPDs obtained by all five methods show that the widths of the marginal PPD for the resistivity and thickness of the second layer are generally wide, indicating large uncertainties in the estimated values of these parameters. The standard deviations of different model parameters (square root of the diagonal element of the posterior model covariance matrix) estimated by the different methods are shown in Figure 8.6. The values estimated by the GS and PGS are almost identical, whereas those obtained by the multiple MAP algorithms are slightly lower than those obtained by the GS and PGS.

The interdependence between different model parameters can be best understood by studying the posterior correlation matrix, which is obtained by normalizing the posterior covariance matrix. A plot of the correlation matrix obtained by each of the five methods is shown in Figure 8.7. Notice that even though the

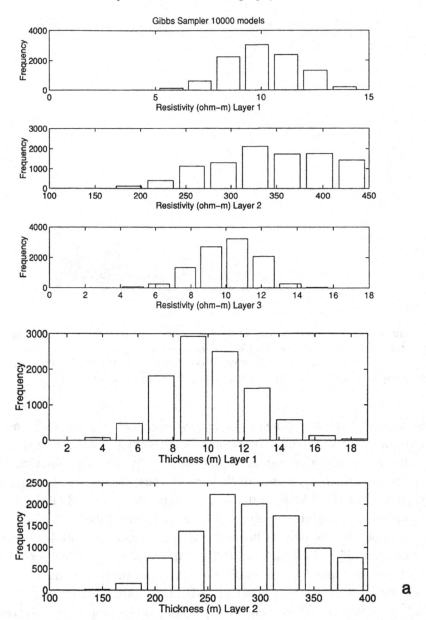

Figure 8.3 Histograms of the model parameters sampled by a Gibbs sampler at $T = 1.0$ after (a) 10,000 and (b) 50,000 model evaluations. Note that histograms at the two iterations look very different. However, beyond iteration 50,000, the histograms did not change significantly, indicating convergence to the stationary distribution.

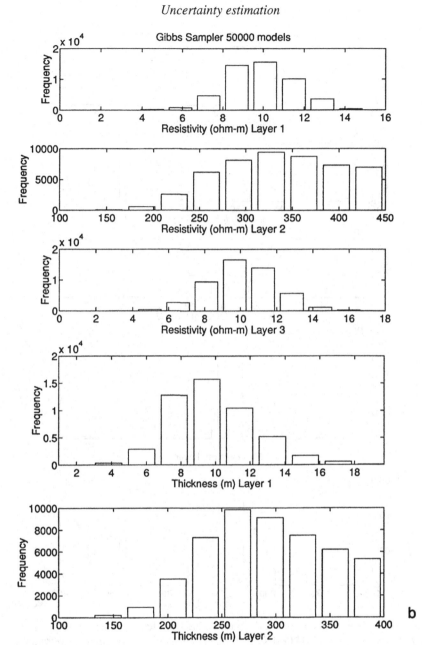

Figure 8.3 (*cont.*)

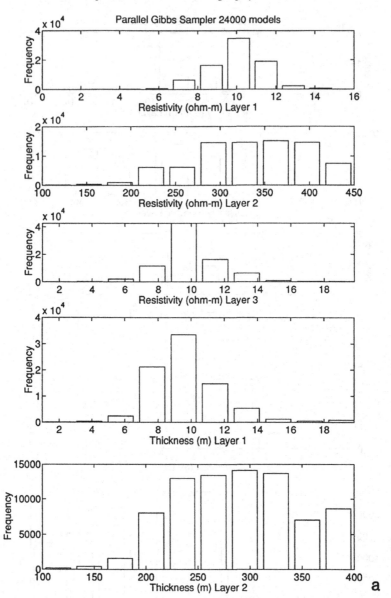

Figure 8.4 Histograms of the model parameters sampled by a parallel Gibbs sampler at $T = 1.0$ after (a) 24,000 and (b) 80,000 iterations. Note that histograms at the two stages look very different. However, beyond 80,000 model evaluations, the histograms did not change any more, indicating convergence to the stationary distribution.

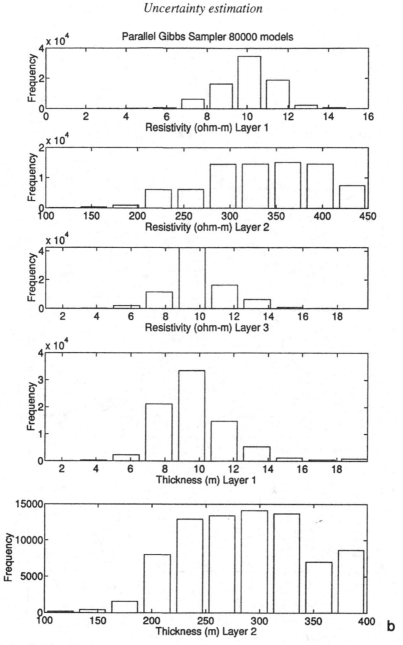

Figure 8.4 (*cont.*)

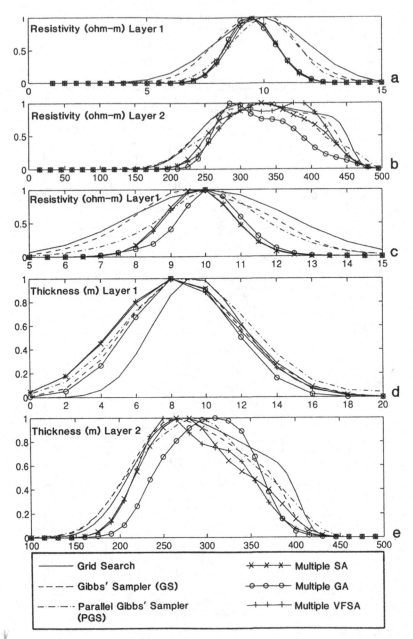

Figure 8.5 Marginal posterior probability density functions (PPD) computed by the grid-search, Gibbs sampler, parallel Gibbs sampler, multiple Metropolis SA, multiple GA, and multiple VFSA for the five model parameters. The curves from all the methods were smoothed using a three-point smoothing operator.

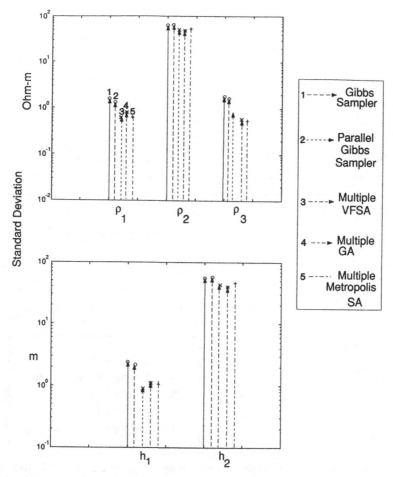

Figure 8.6 Standard deviations of the derived model parameters as estimated by five different algorithms. Note that the GS and PGS estimates are almost identical. However, the standard deviations estimated by multiple MAP algorithms are slightly lower than those estimated by GS and PGS.

multiple MAP algorithms slightly underestimated the covariances, the correlation matrices obtained by all five methods are nearly identical. In particular, notice the strong positive correlation between the resistivity and thickness of the first layer (>0.9). This means that if the resistivity and thickness of the first layer are both increased or decreased, the apparent resistivity curve will not change. This is a classic example of the so-called equivalence problem in vertical electrical sounding interpretation (Koefoed 1979). For this layer, the parameters can be changed (within the variance limits) such that the ratio of the thickness and resistivity, also called the *longitudinal conductance*, remains a constant. We also note that a strong negative correlation is obtained between resistivity and thickness of the second

CORRELATION MATRIX

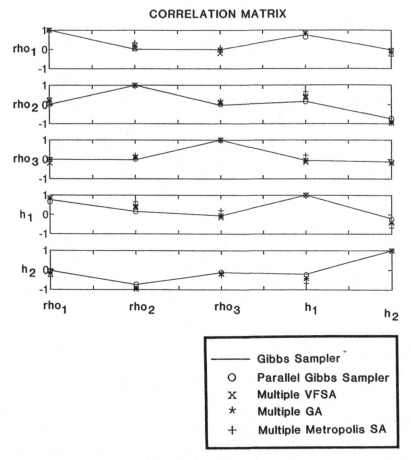

Figure 8.7 Posterior correlation matrix obtained by the five different algorithms. They all are in very good agreement. The strong positive correlation between resistivity and thickness of the first layer and the strong negative correlation between resistivity and thickness of the second layer are well estimated by all the algorithms.

layer. This means that by lowering the resistivity and increasing the thickness, or vice versa, the apparent resistivity curve will not be changed. In this case, it is the *transverse resistance* of the layer given by the product of resistivity and thickness of the layer that remains a constant.

Finally, we compare the computational requirements of different methods by noting the number of forward models, i.e., the number of error function evaluations required, as shown in Figure 8.8. The grid search required the most computation and was computationally intensive, even for a five-parameter problem. The GS and PGS required much less computation time and provided very accurate estimates. The multiple MAP runs were significantly faster than GS and PGS, but the results were not as

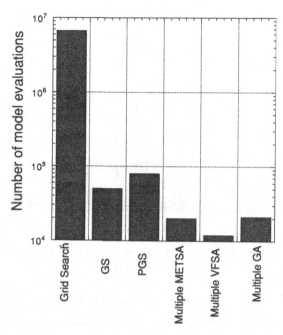

Figure 8.8 Chart showing the number of forward model evaluations required by the different algorithms.

accurate as those obtained by GS and PGS. However, even though the multiple MAP runs underestimated the variances, the correlation values were quite accurate.

8.4.2 Quantifying climate uncertainty

The process of developing an atmospheric general circulation model involves specifying values for a number of uncertain parameters. The parameterization of clouds is thought to be the primary source of uncertainty in predictions of the climate system's response to greenhouse gas forcing. The selection of these parameters can be done through comparison of a number of test experiments that assess which parameter sets allow the model to be most consistent with observational data. The uncertainties and impacts of specifying these parameters in the Community Atmosphere Model (CAM) (Collins *et al.* 2006) were quantified using Bayesian inference and multiple VFSA (MVFSA) stochastic sampling (Jackson *et al.* 2004, 2008). In an updated version of this calculation, Jackson and colleagues consider 15 uncertain model parameters whose effects within CAM were evaluated against observations of seasonal mean surface air temperature, sea level pressure, winds, cloud distributions, surface and top of the atmosphere radiative fluxes, and humidity. Shown in Figure 8.9 are the marginal distributions for six of the most important parameters. Even with as few as 3,281 samples, one can see evidence for distinctly

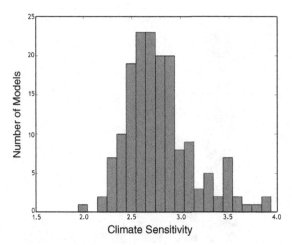

Figure 8.9 Marginal distribution uncertainties for six of fifteen parameters important to clouds within CAM parameters sampled by MVFSA.

non-Gaussian distributions related to model non-linearities. Also shown is the
default parameter setting, indicating that the model configuration selected by hand
through expert analysis is one of the models selected by MVFSA.

Of interest to climate scientists is the equilibrium response of a coupled atmosphere–ocean model to a hypothetical doubling of atmospheric carbon dioxide. This
measure of *climate sensitivity*, as it is commonly called, varies between 1.5°C
and 5°C sensitivity among the approximately 20 models that contribute to the
Intergovernmental Panel on Climate Change (IPCC) assessments of a changing
global environment (e.g., IPCC 2007). Figure 8.10 shows that the uncertainties
in specifying the 15 CAM parameters is associated with a 2°C to 4°C spread in
sensitivity, with 2.7°C occurring most frequently.

This example using MVFSA is the first attempt to apply a sampling approach to
the climate uncertainty problem. It illustrates the practicality of data-driven model
development and quantification of uncertainties for large, complex models with
relatively few forward model integrations.

8.5 Hybrid Monte Carlo

Despite the rigorous proof of convergence of MCMC, its application is often
restricted to relatively small problems because of the large number of forward
model evaluations required for its convergence. Therefore, several variants of
MCMC are being investigated to speed up its convergence (MacKay 2003, ch.
30). Two such algorithms that offer tremendous potential are based on combining
the ideas of local and global search. These are called *Langevin Monte Carlo*, and
hybrid Monte Carlo, or *Hamiltonian Monte Carlo* (HMC) methods.

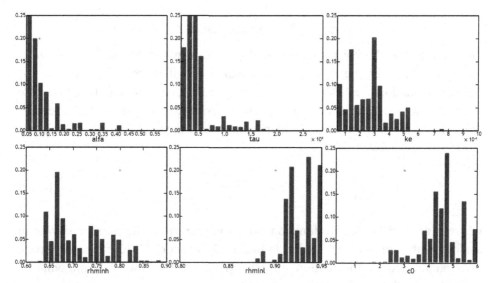

Figure 8.10 The impacts of CAM parameter uncertainties on its global mean air temperature sensitivity to a doubling of carbon dioxide.

8.5.1 Langevin MCMC

You may recall that a gradient-descent algorithm is likely to get trapped in a local minimum if the starting model is too far from the globally optimal solution. The idea behind Langevin MCMC is built on Langevin dynamics (Grenander and Miller 1994); the essential parts of the algorithm can be described in two steps.

In the first step, the model update is computed by adding random noise to the model update predicted by the gradient direction; i.e.,

$$\Delta \mathbf{m} = -\frac{1}{2} \varepsilon^2 \mathbf{g} + \varepsilon \mathbf{n}, \tag{8.9}$$

where $\mathbf{n}$ is a noise vector generated from a Gaussian with unit variance (e.g., MacKay 2003), $\mathbf{g}$ is the vector of gradient of the objective function with respect to model parameters, and ε is a weight term. Note that when ε is zero, Eq. (8.9) reduces to the simple gradient-based model update formula given in Chapter 3.

In the second step of the algorithm, a Metropolis update rule is applied to decide on the acceptance of the model. Thus the trial or proposal distribution of a typical Metropolis–Hastings algorithm is replaced by the gradient direction with an added noise. When used as an optimization approach, this algorithm addresses two fundamental limitations of a gradient-descent algorithm; namely, update is not influenced entirely by the gradient direction, and greediness of a gradient algorithm is replaced by the Metropolis update formula.

8.5.2 Hybrid or Hamiltonian Monte Carlo (HMC)

The Hamiltonian Monte Carlo, originally called a *hybrid Monte Carlo method*, was proposed by Duane *et al.* (1987). It makes use of Hamiltonian dynamics to replace the random-walk Metropolis–Hastings rule. The resulting algorithm, now referred to as a *Hamiltonian Monte Carlo algorithm*, makes use of a momentum vector **p** and the position vector (which in our case is the model vector). Recall that our PPD is given by

$$P(\mathbf{m}) = \frac{e^{-E(\mathbf{m})}}{Z}, \tag{8.10}$$

where $E(\mathbf{m})$ is the objective function, which is also called the *energy function* in analogy with statistical mechanics. In the development of HMC, the model space **m** is augmented by momentum variables **p**, and the following Hamiltonian is defined (e.g., MacKay 2003):

$$H(\mathbf{m}, \mathbf{p}) = E(\mathbf{m}) + K(\mathbf{p}), \tag{8.11}$$

where $K(\mathbf{p})$ is the so-called kinetic energy with

$$K(\mathbf{p}) = \frac{1}{2}\mathbf{p}^T\mathbf{p}. \tag{8.12}$$

Now the joint proposal distribution is given by

$$P_H(\mathbf{m}, \mathbf{p}) = \frac{e^{-H(\mathbf{m},\mathbf{p})}}{Z_H} = \frac{1}{Z_H}e^{-E(\mathbf{m})}e^{-K(\mathbf{p})}. \tag{8.13}$$

Note that since the joint distribution is given by the product of two marginal distributions, the samples of **p** and **m** can be drawn independently. At this stage, we can make use of the well-known Hamiltonian equations to develop an algorithm for time evolution of **m** and **p** (e.g., MacKay 2003). A flowchart of the resulting algorithm is shown in Figure 8.11.

The success of both Langevin MCMC and HMC depends on the efficiency of computation of the gradient of the objective function. In many applications, such as the inversion of electromagnetic data and seismic waveform data, the gradient of the objective function can be computed by an adjoint state approach, which requires only one extra forward calculation. Thus, for these applications, these two algorithms offer tremendous potential.

Draw a model **m** from prior distribution and compute the objective function $E(\mathbf{m})$ and

gradient vector $g(\mathbf{m}) = \dfrac{\partial E}{\partial \mathbf{m}}$

Loop i

    draw initial **p** from N(0,1)

    $H = \dfrac{\mathbf{p}^T \mathbf{p}}{2} + E$

    $\mathbf{m}^{new} = \mathbf{m}$; $\mathbf{g}^{new} = \mathbf{g}$

    Loop j

        $\mathbf{p} = \mathbf{p} - \mathbf{g}^{new} / 2$

        $\mathbf{m}^{new} = \mathbf{m}^{new} + \mathbf{p}$

        $\mathbf{g}^{new} = \left. \dfrac{\partial E}{\partial \mathbf{m}} \right|_{\mathbf{m}=\mathbf{m}^{new}}$

        $\mathbf{p} = \mathbf{p} - \dfrac{1}{2} \mathbf{g}^{new}$

    end loop j

        Compute E^{new}

        $H^{new} = \dfrac{1}{2} \mathbf{p}^T \mathbf{p} + E^{new}$

        $\Delta H = H^{new} - H$

        Apply Metropolis rule over ΔH

    end of loop i

Figure 8.11 A pseudocode of the Hamiltonian Monte Carlo algorithm. (Modified from MacKay 2003.)

8.6 Summary

It is now well recognized that the results of geophysical inversion may have uncertainties for several reasons. Bayes' rule offers an elegant description of the solution for the geophysical inverse problem in terms of the posterior probability density function in the model space. Because the model space is highly multidimensional, a complete display of the PPD is not possible, and therefore, we seek quantities such as the posterior marginal density functions, mean, covariance, etc. to describe the solution of our inverse problems. Derivation of these quantities is not a trivial task because such derivations involve numerical evaluation of integrals with a large number of variables. One additional difficulty is that for most geophysical inverse problems, the data have a non-linear relationship with the model parameters, resulting in error functions that may have multiple extrema. The PPD therefore may be highly complicated, and methods that require prior assumptions about the shape of the PPD may not be adequate. These methods attempt to locate the maximum of the PPD, and the Hessian computed at the MAP point is used to estimate the covariance; thus a Gaussian is completely described.

We described several computationally intensive, sampling-based approaches to derive quantities such as the posterior marginal PPD, mean, and covariance. Six different methods have been tested and compared by application to synthetic resistivity sounding data where the number of model parameters was only five, such that the forward modeling was computationally tractable. Methods such as GS and PGS agree well with grid-search integration. The results from multiple MAP algorithms are quite encouraging because even though the estimated variances were lower than those estimated by GS, PGS, and grid search, the correlations were in very good agreement. Of course, for very complicated PPDs, the multiple MAP algorithms may not perform equally well, but for many applications, they will help in getting approximate estimates of these quantities quite rapidly.

The SA method is guaranteed to asymptotically converge to a Gibbs distribution. We showed numerically that a steady-state distribution can be achieved in a finite number of iterations. Thus a GS requires much less computation time than a complete grid search in the evaluation of multidimensional integrals. Running SA at a constant temperature of 1.0 appears to be feasible for many practical problems. Although our PGS is not based on a mathematical model as rigorous as SA, we showed numerically that the results from GS and PGS are nearly identical. The PGS, however, offers the advantage of running the algorithm on parallel computer architectures.

Note that the PGS based on multiple SA or GAs does not satisfy the properties of the Markov chain and is therefore not a true MCMC method. Nonetheless, this algorithm is found to be extremely efficient, and the uncertainty estimates are fairly general. The idea of making parallel runs is novel, and interestingly, algorithms called *simulated tempering* and *parallel tempering* have also been developed (e.g., Gregory 2005, ch. 12). The principal idea behind parallel tempering is to run multiple parallel chains at a constant temperature each. At each step, samples are swapped using the Metropolis criterion. At the end, we expect the samples from the chain at $T = 1$ to be converging to our PPD.

The use of stochastic inversion methods based on statistical models has been criticized primarily because of the requirement that the prior distribution of noise in the data and theory error be known (Tarantola 1987; Scales and Snieder 1997; Scales and Tenerio 2001).

Often the prior noise pdf is assumed to be Gaussian. In practice, it is never possible to estimate the true noise distribution with a small and finite number of repeated measurements because the classical definition of probability requires that an experiment be repeated an infinite number of times under identical conditions. The use and specification of the prior pdf have long been a subject of dispute, and the use of a statistical approach often has been criticized by classical inverse theorists. There exist several definitions of probability. Besides the classical definition, which is based on relative frequency of occurrence, the Bayesian interpretation defines probability as a *degree of belief* (Bayes 1763). The only requirement is that the axioms of probability theory are not violated. Again, in the subjective Bayesian interpretation, the *degree of belief* is a personal degree of belief. Thus a priori distribution may simply reflect an expert's opinion. We therefore recommend that a simple and practical approach be taken to characterize the prior distribution of noise in the data. For any practical problem, we need to carefully analyze the data and explore the sources of error in the data in order to arrive at some way of characterizing the uncertainty.

Bibliography

Aarts, E., and Korst, J., 1989, *Simulated Annealing and Boltzmann Machines*, John Wiley and Sons, New York.

Abramowitz, M., and Stegun, I. A., 1972, *Handbook of Mathematical Functions*, Dover Publications, New York.

Aki, K., and Richards, P. G., 1980, *Quantitative Seismology*, Vols. I and II, Freeman and Co., San Francisco.

Ammon, C. J., and Vidale, J. E., 1993, Tomography without rays, *Bull. Seism. Soc. Am.* 83, 509–528.

Anily, S., and Federgruen, A., 1987a, Simulated annealing methods with general acceptance probabilities, *J. Appl. Prob.* 24, 657–667.

Anily, S., and Federgruen, A., 1987b, Ergodicity in non-stationary Markov chains: An application to simulated annealing methods, *Operations Research* 35, 867–874.

Aster, R. C., Brochers, B., and Thurber, C. H., 2005, *Parameter Estimation and Inverse Problems*, Elsevier Academic Press, New York.

Backus, G. E., 1970, Inference from inadequate and inaccurate data, I, *Proc. Natl. Acad. Sci. USA* 65, 1–7.

Backus, G. E., and Gilbert, F., 1967, Numerical applications of a formalism for geophysical inverse problems, *Geophys. J. R. Astr. Soc.* 13, 247–276.

Backus, G. E., and Gilbert, F., 1968, The resolving power of gross earth data, *Geophys. J. R. Astr. Soc.* 16, 169–205.

Backus, G. E., and Gilbert, F., 1970, Uniqueness in the inversion of inaccurate gross earth data, *Phil. Trans. R. Soc. Lond. A.* 266, 123–192.

Baker, J. E., 1987, Reducing bias and inefficiency in the selection algorithm, in J. J. Grefenstette (Ed.), *Proceedings of the Second International Conference on Genetic Algorithms*, 14–21.

Bangerth, W., Klie, H., Wheeler, M. F., Stoffa, P. L., and Sen, M. K., 2006, On optimization algorithms for the reservoir oil well placement problem, *Comput. Geosci.* 10, 303–319.

Barber, D., and Williams, C. K. I., 1997, Gaussian processes for Bayesian classification via hybrid Monte Carlo, in M. C. Mozer, M. I. Jordan, and T. Petsche (Eds.), *Neural Information Processing Systems*, Vol. 9, MIT Press, Boston, pp. 340–346.

Bard, Y., 1974, *Nonlinear Parameter Estimation*, Academic Press, New York.

Basu, A., and Frazer, L. N., 1990, Rapid determination of critical temperature in simulated annealing inversion, *Science* 249, 1409–1412.

Bayes, T., 1763, Essay towards solving a problem in the doctrine of chances; republished in 1958 in *Biometrika* 45, 293–315.

Belge, M., Kilmer, M. E., and Miller, E. L., 1998, Simultaneous multiple regularization parameters selection by means of *L*-hypersurface with application to linear inverse problems posed in the wavelet transform domain, *Proc. SPIE* 3459, 328–336.

Berg, E., 1990, Simple convergent genetic algorithms for inversion of multiparameter data, *Proceedings of the Society of Exploration Geophysicists Sixtieth Annual International Meeting and Exposition*, Expanded Abstracts, Vol. II, 1126–1128.

Bhattacharya, B. B., Sen, M. K., Stoffa, P. L., GhoshRoy, I., and Mohanty, P. R., 1992, Global optimization methods in the interpretation of VES and MTS data (expanded abstract), paper presented at the International Conference on Mathematical Modeling and Computer Simulation, Bangalore, India.

Bhattacharya, P. K., and Patra, H. P., 1968, *Direct Current Geoelectric Sounding: Principles and Interpretation*, Elsevier Science Publishers, New York.

Box, G. P., and Tiao, G. C., 1973, *Bayesian Inference in Statistical Analysis*, Addison-Wesley, Reading, MA.

Brysk, H., and McCowan, D. W., 1986, Edge effects in cylindrical slant stacks, *Geophys. J. R. Astr. Soc.* 87, 801–813.

Buland, A., and Omre, H., 2003, Bayesian linearized AVO inversion, *Geophysics* 68(1), 185–198.

Cagniard, L., 1953, Basic theory of magnetotelluric method of geophysical prospecting, *Geophysics* 18, 605–635.

Calderon, C., and Sen, M. K., 1995, Seismic deconvolution by mean field annealing, paper presented at the SIAM Conference on Mathematical and Computational Issues in Geosciences, San Antonio, TX.

Cary, P. W., and Chapman, C. H., 1988, Automatic 1D waveform inversion of marine seismic reflection data, *Geophys. J. Int.* 93, 527–546.

Chunduru, R. K., Sen, M. K., and Stoffa, P. L., 1994, Resistivity inversion for 2D geologic structures using very fast simulated annealing, *Proceedings of the Society of Exploration Geophysicists Sixty-Fourth Annual International Meeting and Exposition*, Los Angeles, CA, 640–643.

Chunduru, R. K., Sen, M. K., Stoffa, P. L., and Nagendra, R., 1995, Non-linear inversion of resistivity profiling data for some geometrical bodies, *Geophysical Prospecting* 43, 979–1003.

Chunduru, R. K., Sen, M. K., and Stoffa, P. L., 1997, Hybrid optimization methods for geophysical inversion, *Geophysics* 62(4), 1196–1207.

Cichocki, A., and Unbejauen, R., 1993, *Neural Networks for Optimization and Signal Processing*, John Wiley and Sons, New York.

Claerbout, J. F., and Muir, F., 1973, Robust modeling with erratic data, *Geophysics* 38, 826–844.

Clarke, T. J., 1984, Full reconstruction of a layered elastic medium from P-SV slant stack data, *Geophys. J. R. Astr. Soc.* 78, 775–793.

Clayton, R. W., and Stolt, R. H., 1981, A Born–WKBJ inversion method for acoustic reflection data, *Geophysics* 46, 1559–1567.

Cohoon, J. P., Hegde, S. U., Martin, W. N., and Richards, S., 1987, Punctuated equilibria: A parallel genetic algorithms, in J. J. Grefenstette (Ed.), *Proceedings of the Second International Conference on Genetic Algorithms*, 148–154.

Collins, W. D., Rasch, P. J., Boville, B. A., Hack, J. J., McCaa, J. R., Williamson, D. L., Briegleb, B. P., Bitz, C. M., Lin, S. J., and Zhang, M., 2006, The formulation and atmospheric simulation of the community atmosphere model version 3 (CAM3), *J. Climate* 19, 2144–2161.

Craven P., and Wahba G., 1979, Smoothing noisy data with spline functions: Estimating correct degree of smoothing by the method of generalized cross-validation: Numerische Mathematik, 31, 377–403.

Creutz, M., 1984, *Quarks, Gluons and Lattices*, Cambridge University Press, Cambridge, UK.

Curtis, A., and Lomax, A., 2001, Prior information, sampling distributions and the curse of dimensionality, *Geophysics* 66, 372–378.

Davis, L. D., 1991, *Handbook of Genetic Algorithms*, Van Nostrand Reinhold, New York.

Davis, L., and Coombs, S., 1987, Genetic algorithms and communication link speed design: Theoretical considerations, in J. J. Grefenstette (Ed.), *Proceedings of the Second International Conference on Genetic Algorithms*, 252–256.

Davis, T. E., and Principe, J. C., 1991, A simulated annealing-like convergence theory for the simple genetic algorithm, in R. K. Belew and L. B. Booker (Eds.), *Proceedings of the Fourth International Conference on Genetic Algorithms*, 174–181.

Deb, K., and Goldberg, D. E., 1989, An investigation of niche and species formation in genetic function optimization, in J. D. Schaffer (Ed.), *Proceedings of Third International Conference on Genetic Algorithms*, 42–50.

Deeming, T. J., 1987, Band-limited minimum phase, in deconvolution and inversion, in *Proceedings of the 1986 Rome Workshop on Deconvolution and Inversion*, Blackwell Scientific Publishers, Boston.

Deutsch, C. V., and Journel, A. G., 1991, The application of simulated annealing to stochastic reservoir modeling, Paper SPE 23565, Society of Petroleum Engineers.

Dey, A., and Morisson, H. F., 1979, Resistivity modeling for arbitrarily shaped three-dimensional structures, *Geophysics* 44, 753–780.

Dimri, V. P., and R. P. Srivastava, 2005, Fractal modeling of complex subsurface geological structures, in V. P. Dimri (Ed.), *Fractal Behaviour of the Earth System*, Springer, New York, pp. 23–37.

Dimri, V. P., R. P. Srivastava, and N. Vedanti, 2012, *Fractal models in exploration geophysics*, Elsevier publications, The Netherlands.

Dorigo, M., 1992, Learning and natural algorithms, PhD thesis, Politecnico di Milano, Italy.

Dorigo, M., and Stutzle, T., 2004, *Ant Colony Optimization*, MIT Press, Boston.

Dosso, S. E., and Oldenburg, D. W., 1991, Magnetotelluric appraisal using simulated annealing, *Geophys. J. Int.* 106, 379–385.

Duane, S., Kennedy, A. D., Peddleton, B. J., and Roweth, D., 1987, Hybrid Monte Carlo, Physics Letters B, 195: 216–222.

Duijndam, J. W., 1987, Detailed Bayesian inversion of seismic data, PhD dissertation, Delft University of Technology, The Netherlands.

Duijndam, J. W., 1988a, Bayesian estimation in seismic inversion, I: Principles, *Geophys. Prosp.* 36, 878–898.

Duijndam, J. W., 1988b, Bayesian estimation in seismic inversion, II: Uncertainty analysis, *Geophys. Prosp.* 36, 899–918.

Dziewonski, A. M., 1984, Mapping the lower mantle: Determination of lateral heterogeneity in *P* velocity up to degree and order 6, *J. Geophys. Res.* 89, 5929–5952.

Eiben, A. E., Aarts, E., and Van Hee, K. M., 1991, Global convergence of genetic algorithms: A Markov chain analysis, in H. P. Schwefel and R. Männer (Eds.), *Parallel Problem Solving from Nature*, Springer-Verlag, New York, pp. 4–12.

Eldredge, N., and Gould, S. J., 1972, Punctuated equilibria: An application to phyletic gradualism, in T. J. M. Schopf (Ed.), *Models in Paleobiology*, Freeman, Cooper, San Francisco, pp. 82–115.

Engl, H. W., 1987, Discrepancy principles of Tikhonov regularization of ill-posed problems leading to optimal convergence rates, *J. Optim. Theo. Appl.* 52, 209–215.

Farmer, C. L., 1988, The generation of stochastic fields of reservoir parameters with specified geostatistical distributions, in S. S. Edwards and P. King (Eds.), *Mathematics of Oil Production*, Clarendon Press, London, pp. 235–252.

Feller, W., 1968, *An Introduction to Probability Theory and Its Applications*, John Wesley and Sons, London.

Fernández-Alvarez, J. P., Fernández-Martínez, J. L., and Menéndez Pérez, C.O., 2008, Feasibility of the use of binary genetic algorithms as importance samplers: Application to a 1D-DC inverse problem, *Math. Geosci.* DOI 10.1007/s11004-008-9151-y.

Fernández-Martínez, J. L., and García-Gonzalo, E., 2008a, The generalized PSO: A new door to PSO evolution, *J. Artif. Evol. Appl.* DOI:10.1155/2008/861275.

Fernández-Martínez, J. L., García-Gonzalo, E., and Fernández-Alvarez, J. P., 2008b, Theoretical analysis of particle swarm trajectories through a mechanical analogy, *Int. J. Comput. Intell. Res.*, Special Issue on PSO, 2008.

Fogel, L. J., Owens, A. J., and Walsh, M. J., 1966, *Artificial Intelligence Through Simulated Evolution*, John Wiley and Sons, New York.

Fogel, L. J., and Fogel, D. B., 1986, Artificial intelligence through evolutionary programming, final report to ARI, contract no. PO–9–X56–1102C–1.

Fogel, D. B., 1988, An evolutionary approach to traveling salesman problem, *Biol. Cybernet.* 60, 139–144.

Fogel, D. B., 1991, *System Identification Through Simulated Evolution: A Machine Learning Approach to Modeling*, Ginn Press, Boston.

Forrest, Stephanie, 1993, Genetic algorithms: Principles of natural selection applied to computation, *Science* 261, 872–878.

Franklin, J. N., 1970, Well-posed stochastic extensions of ill-posed linear problems, *J. Math. Anal. Appl.* 31, 682–716.

Friedberg, R. M., 1958, A learning machine: Part I, *IBM J. Res. Dev.* 2, 2–13.

Fuchs, K., and Müller, G., 1971, Computation of synthetic seismograms with reflectivity method and comparison with observations, *Geophys. J. Roy. Astr. Soc.* 23, 417–433.

Gelfand, A. E., and Smith, A. F. M., 1990, Sampling based approaches to calculating marginal densities, *J. Am. Stat. Assoc.* 85, 398–409.

Gelfand, A. E., Smith, A. F. M., and Lee, Tai-Ming, 1992, Bayesian analysis of constrained parameter and truncated data problems using Gibbs' sampling, *J. Am. Stat. Assoc.* 87, 523–532.

Geman, S., and Geman, D., 1984, Stochastic relaxation, Gibbs' distribution and Bayesian restoration of images, *IEEE Trans.* PAMI-6, 721–741.

Gilks, W. R., and Wild, P., 1992, Adaptive rejection sampling for Gibbs' sampling, *Appl. Stat.* 41, 337–348.

Gill, P. E., Murray, W., and Wright, M. H., 1981, *Practical Optimization*, Academic Press, New York.

Girard, D., 1989, A fast Monte-Carlo cross-validation procedure for large least squares problems with noisy data, *Numerische Mathematik* 56, 1–23.

Glover, D. E., 1987, Solving a complex keyboard configuration problem through generalized adaptive search, in L. Davis (Ed.), *Genetic Algorithms and Simulated Annealing*, Morgan Kaufmann, Los Altos, CA, pp. 12–31.

Goldberg, D. E., and Richardson, J., 1987, Genetic algorithms with sharing for multimodal function optimization, in J. J. Grefenstette (Ed.), *Proceedings of the Second International Conference on Genetic Algorithms*, 41–49.

Goldberg, D. E., and Segrest, P., 1987, Finite Markov chain analysis of genetic algorithms, in J. J. Grefenstette (Ed.), *Proceedings of the Second International Conference on Genetic Algorithms*, 1–8.

Goldberg, D. E., 1989, *Genetic Algorithms in Search, Optimization and Machine Learning*, Addison Wesley, Reading, MA.

Goldberg, D. E., and Deb, K., 1991, A comparative analysis of selection schemes used in genetic algorithms, in G. J. E. Rawlins (Ed.), *Foundations of Genetic Algorithms*, Morgan Kaufmann, San Mateo, CA, pp. 69–93.

272 *Bibliography*

Greene, J. W., and Supowit, K. J., 1986, Simulated annealing without rejected moves, *IEEE Trans. Computer-Aided Design* CAD-5(1), 221–228.

Gregory, P., 2005, *Bayesian Logical Data Analysis for the Physical Sciences*, Cambridge University Press, Cambridge, UK.

Grenander, U., and Miller, M. I., 1994, Representations of knowledge in complex systems, *J. R. Stats. Soc. B* 56(4), 549–603.

Haldorsen, H., and Damsleth, E., 1990, Stochastic modeling, *JPT* (April). 42, 404–412.

Hammersley, J. M., and Handscomb, D. C., 1964, *Monte Carlo Methods*, Chapman & Hall, London.

Hampson, D., 1991, AVO inversion, theory and practice, *The Leading Edge* 10, 39–42.

Hanke, M., 1997, Regularizing properties of a truncated Newton-CG algorithm for nonlinear inverse problems, *Numer. Funct. Anal. Optim.* 18, 971–993.

Hansen, P. C., 1992, Analysis of discrete ill-posed problems by means of L-curve, *SIAM Rev.* 34, 561–580.

Hansen, P. C., 2001, The L-curve and its use in the numerical treatment of inverse problems, in P. Johnston (Ed.), *Computational Inverse Problems in Electrocardiology*, MIT Press, Boston, pp. 119–142.

Hansen, P.C., 2007, Regularization tools version 4.0 Matlab 7.3, *Numer. Alogor.*, 46, 189–194.

Hansen, P. C., and O'Leary, D. P., 1993, The use of L-curve in the regularization of discrete ill-posed problems, *SIAM J. Sci. Comp.* 14, 1487–1503.

Hastings, W. K., 1970, Monte Carlo methods using Markov chains and their applications, *Biometrika* 57, 97–109.

Hewett, T., 1986, Fractal distribution of reservoir heterogeneity and their influence on fluid transport, paper SPE 15386, presented at the Annual Technical Conference and Exhibition, New Orleans, LA.

Hinton, G. F., Sejnowski, T. J., and Ackley, D. H., 1984, Boltzmann machines: Constrained satisfaction networks that learn, Carnegie Mellon University technical report, CMU–CS–84–119.

Holland, J. H., 1975, *Adaptation in Natural and Artificial Systems*, University of Michigan Press, Ann Arbor.

Hopfield, J. J., 1982, Neural networks and physical systems with emergent collective computational properties, *Proc. Natl. Acad. Sci. USA* 79, 2554–2558.

Hopfield, J. J., 1984, Neurons with graded response have collective computation properties like those of two-state neurons, *Proc. Natl. Acad. Sci. USA* 81, 3088–3092.

Hopfield, J. J., and Tank, D. W., 1985, Neural computation of decisions in optimization problems, *Biol. Cybernet.* 52, 141–152.

Hopfield, J. J., and Tank, D. W., 1986, Computing with neural circuits: A model, *Science* 233, 625–633.

Horn, J., 1993, *Finite Markov chain analysis of genetic algorithms with niching*, Illigal Report No. 93002, University of Illinois at Urbana-Champaign.

Hutchinson M. F., 1989, A stochastic estimator of the trace of the influence matrix of Laplacian smoothing splines: Comm. Stat. Simul. Comp., 18, 1059–1076.

Ingber, L., 1989, Very fast simulated reannealing, *Mathl. Comput. Modeling* 12(8), 967–993.

Ingber, L., 1993, Simulated annealing: Practice versus theory, *Mathl. Comput. Modeling* 18(11), 29–57.

Ingber, L., and Rosen, B., 1992, Genetic algorithms and simulated reannealing: A comparison, *Mathl. Comput. Modeling* 16(11), 87–100.

IPCC, 2007, *Climate Change 2007: The Physical Science Basis. Contribution of Working Group I to the Fourth Assessment Report of the Intergovernmental Panel on Climate*

Change [Solomon, S., Qin, D., Manning, M., Chen, Z., Marquis, M., Averyt, K. B., Tignor, M., and Miller, H. L., Eds.], Cambridge University Press, Cambridge, UK.

Jackson, C., Sen, M., and Stoffa, P., 2004, An efficient stochastic Bayesian approach to optimal parameter and uncertainty estimation for climate model predictions, *J. Climate* 17(14), 2828–2841.

Jackson, C. S., Sen, M. K., Huerta, G., Deng, Y., and Bowman, K. P., 2008, Error reduction and convergence in climate prediction, *J. Climate* 21(24), 6698–6709. DOI: 10.1175/2008JCLI2112.1.

Jackson, D. D., 1972, Interpretation of inaccurate, insufficient and inconsistent data, *Geophys. J. R. Astr. Soc.* 28(2), 97–109.

Jackson, D. D., 1979, The use of a priori data to resolve non-uniqueness in linear inversion, *Geophys. J. R. Astr. Soc.* 57, 137–157.

Jackson, D. D., and Matsura, M., 1985, A Bayesian approach to nonlinear inversion, *J. Geophys. Res.* 90(B1), 581–591.

Jeffreys, H., 1939, *Theory of Probability*, Clarendon Press, London.

Jervis, M., 1993, Optimization methods for 2D pre-stack migration velocity estimation, PhD dissertation, The University of Texas, Austin.

Jervis, M., Sen, M. K., and Stoffa, P. L., 1993a, 2D migration velocity estimation using a genetic algorithm, *Geophys. Res. Lett.* 20, 1495–1498.

Jervis, M., Sen, M. K., and Stoffa, P. L., 1993b, Optimization methods for 2D pre-stack migration velocity estimation, in *Proceedings of the Society of Exploration Geophysicists Sixty-Third Annual International Meeting and Exposition*, 965–968.

Jin, S., and Madariaga, R., 1993, Background velocity inversion with a genetic algorithm, *Geophys. Res. Lett.* 20, 93–96.

Jin, L., Stoffa, P. L., Sen, M. K., Seif, R. K., and Sena, A., 2009, Pilot point parameterization in stochastic inversion for reservoir properties using time-lapse seismic and production data, *J Seis. Explor.* 18(1), 1–20.

Journel, A., and Alabert, F., 1990, New method of reservoir mapping, *JPT*, 42(2), 212–218.

Kennedy, J. and Eberhart, R. C. 1995, Particle swarm optimization. Proceedings of IEEE International Conference on Neural Networks, Piscataway, NJ. pp. 1942–1948.

Kemeny, J. G., and Snell, J. L., 1960, *Finite Markov chains*, D. Van Nostrand, New York.

Kennedy, W. J., Jr., and Gentle, J. F., 1980, *Statistical Computing*, Marcel Dekker, New York.

Kennett, B. L. N., 1983, *Seismic Wave Propagation in Stratified Media*, Cambridge University Press.

Kennett, B. L. N., and Nolet, G., 1978, Resolution analysis for discrete systems, *Geophys. J. R. Astr. Soc.* 53, 413–425.

Kennett, B. L. N., and Sambridge, M. S., 1992, Earthquake location – genetic algorithms for teleseisms, *Phys. Earth Planet. Int.* 75, 103–110.

Kirkpatrick, S., Gelatt, C. D., Jr., and Vecchi, M. P., 1983, Optimization by simulated annealing, *Science* 220, 671–680.

Koch, M., 1990, Optimal regularization of the linear seismic inverse problem, Technical Report No. FSU-SCRI-90C-32, Florida State University, Tallahassee.

Koefoed, O., 1979, *Geosounding Principles, 1, Resistivity Sounding Measurements*, Elsevier Science Publishers, New York.

Koza, J. R., 1992, *Genetic Programming*, MIT Press, Boston.

Lanczos, C., 1961, *Differential Operators*, D. Van Nostrand, London.

Landa, E., Beydoun, W., and Tarantola, A., 1989, Reference velocity model estimation from prestack inversion: Coherency optimization by simulated annealing, *Geophysics* 54, 984–990.

Levy, S., Oldenburg, D., and Wang, J., 1988, Subsurface imaging using magnetotelluric data, *Geophysics* 53, 104–117.

Lines, L. R., and Treitel, S., 1984, Tutorial: A review of least-squares inversion and its application to geophysical problems, *Geophys. Prosp.* 32, 159–186.

MacKay, D. J. C., 2003, *Information Theory, Inference and Learning Algorithms*, Cambridge University Press, Cambridge, UK.

Mahfoud, S. W., 1991, Finite Markov chain models of an alternative selection strategy for the genetic algorithm, Illigal Report No. 91007, University of Illinois at Urbana-Champaign.

Malinverno, A., 2000. A Bayesian criterion for simplicity in inverse problem parametrization, *Geophys. J. Int.* 140, 267–285.

Malinverno, A., 2002, Parsimonious Bayesian Markov chain Monte Carlo inversion in a nonlinear geophysical problem, *Geophysical J. Int.* 151(3), 675–688.

Malinverno, A., and Briggs, V.A., 2004, Expanded uncertainty quantification in inverse problems: Hierarchical Bayes and empirical Bayes, *Geophysics* 69, 1005–1016.

Malinverno, A., and Leaney, W. S., 2005, Monte Carlo Bayesian look-ahead inversion of walkaway vertical seismic profiles, *Geophys. Prospect.* 53, 689–703.

Mallick, S., 1999, Some practical aspects of prestack waveform inversion using a genetic algorithm: An example from the East Texas Woodbine gas sand, *Geophysics* 64, 326–336.

Matlab, manual, MathWorks, Inc., 3 Apple Hill Drive, Natick, MA 01760–2098.

McAulay, A. D., 1985, Prestack inversion with plane-layer point source modeling, *Geophysics* 50, 77–89.

Menke, W., 1984, *Geophysical Data Analysis: Discrete Inverse Theory*, Academic Press, New York.

Metropolis, N., and Ulam, S., 1949, The Monte Carlo method, *J. Acous. Soc. Am.* 44, 335–341.

Metropolis, N., Rosenbluth, A., Rosenbluth, M., Teller, A., and Teller, E., 1953, Equation of state calculations by fast computing machines, *J. Chem. Phys.* 21, 1087–1092.

Mitra, D., Ramio, F., and Sangiovanni-Vincentelli, A., 1986, Convergence and finite time behavior of simulated annealing, *Adv. Appl. Prob.*, 747–771.

Morozov, V. A., 1986, *Methods for Solving Incorrectly Posed Problems*, Springer-Verlag, New York.

Mosegaard, K., and Sambridge, M., 2002. Monte Carlo analysis of inverse problems, *Inverse Problems* 18, R29–R54.

Mosegaard, K., and Tarantola, A., 1995, Monte Carlo sampling of solutions to inverse problems, *J. Geophys. Res.* 100, 12431–12447.

Mufti, I. R., 1976, Finite-difference resistivity modeling for arbitrarily shaped two-dimensional structures, *Geophysics* 41, 62–78.

Narayan, S., and Dusseault, M. B., 1992, Resistivity inversion method applied to shallow structural problems, *Geotechnique et Informatique*, October.

Neumaier A., 1998, Solving ill-conditioned and singular linear systems – A tutorial on regularization: SIAM Review, 40, 636–666.

Nix, A. E., and Vose, M. D., 1992, Modeling genetic algorithms with Markov chains, *Ann. Math. Artif. Intell.* 5, 79–88.

Ooe, M., and Ulrych, T. J., 1979, Minimum entropy deconvolution with an exponential transform, *Geophys. Prosp.* 27, 458–473.

Okabe, A., Boots, B., Sahigara, K., and Chiu, S. N., 1999, *Spatial Tessallations: Concepts and Applications of Voronoi Diagrams*, John Wiley and Sons, New York.

Oldenburg, D. W., 1979, One-dimensional inversion of natural source magnetotelluric observations, *Geophysics* 44, 1218–1244.

Papoulis, A., 1965, *Probability, Random Variables and Stochastic Processes*, McGraw-Hill, New York.

Parker, R. L., 1972, Inverse theory with grossly inadequate data, *Geophys. J. R. Astr. Soc.* 29, 123–128.

Parker, R. L., 1977, Understanding inverse theory, *Ann. Rev. Earth Planet. Sci.* 5, 35–64.

Peterson, C., and Anderson, J. R., 1987, A mean field theory learning algorithm for neural networks, *Complex Systems* 1, 995–1019.

Peterson, C., and Anderson, J. R., 1988, Neural networks and NP-complete optimization problems: A performance study on the graph bisection problem, *Complex Systems* 2, 59–89.

Peterson, C., and Soderberg, B., 1989, A new method for mapping optimization problems onto neural networks, *Int. J. Neural Sci.* 1, 3–22.

Porsani, M. J., Stoffa, P. L., Sen, M. K., Chunduru, R. K., and Wood, W. T., 1993, A combined genetic and linear inversion algorithm for seismic waveform inversion, in *Proceedings of the Society of Exploration Geophysicists Sixty-Third Annual International Meeting and Exposition*, Washington, DC, 692–695.

Porstendorfer, G., 1975, *Principles of Magnetotelluric Prospecting*, Geoexploration Monograph Series 1, No. 5, Gebruder Borntraeger, Berlin.

Press, F., 1968, Earth models obtained by Monte-Carlo inversion, *J. Geophys. Res.* 73, 5223–5234.

Press, W. H., Flannerry, B. P., Teukolsky, S. A., and Vetterling, W. T., 1989, *Numerical Recipes: The Art of Scientific Computing*, Cambridge University Press, Cambridge, UK.

Pullammanappallil, S. K., and Louie, J. N., 1993, Inversion of reflection travel time using a nonlinear optimization scheme, *Geophysics* 58, 1607–1620.

Ramlau, R., 2002, A steepest descent algorithm for the global minimization of Tikhonov functional, *Inverse Problems* 18, 381–403.

Rebbi, C., 1984, Monte Carlo calculations in lattice gauge theories, in K. Binder (Ed.), *Applications of the Monte Carlo Method*, Springer-Verlag, New York, pp. 277–298.

Reginska, T., 1996, A regularization parameter in discrete ill-posed problems, *SIAM J. Sci. Comp.* 17, 740–749.

Ross, S. M., 1989, *Introduction to Probability Models*, Academic Press, New York.

Rothman, D. H., 1985, Nonlinear inversion, statistical mechanics, and residual statics estimation, *Geophysics* 50, 2784–2796.

Rothman, D. H., 1986, Automatic estimation of large residual statics corrections, *Geophysics* 51, 337–346.

Roy, L., Sen, M. K., Stoffa, P. L., McIntosh, K., and Nakamura, Y., 2005, Joint inversion of first arrival travel time and gravity data, *J. Geophys. Eng.* 2, 277–289.

Rubinstein, R. Y., 1981, *Simulation and the Monte Carlo Method*, John Wiley and Sons, New York.

Rudin, Leonid I., Osher, Stanley, and Fatemi, Emad, 1992, Nonlinear total variation based noise removal algorithms, in *Proceeding of the 11th Annual International Conference of the Center for Nonlinear Studies, Physica D* 60, 259–268.

Rumelhart, D. E., Hinton, G. E., and Williams, R. J., 1986, Learning internal representations by error propagation, in D. E. Rumelhart and J. L. McClelland (Eds.), *Parallel Distributed Processing: Exploration in the Microstructure of Cognition*, Vol. 1: *Foundations*, MIT Press, Boston, 318–362.

Sadegh, P., and Spall, J. C., 1996, Optimal sensor configuration for complex systems, *Proc. Test Technol. Symp.* (sponsored by U.S. Army TECOM); available at: www.atc.army.mil/-tecom/tts/proceed/optsenr.html.

Sambridge, M., 1999a, Geophysical inversion using a neighborhood algorithm. I. Searching a parameter space, *Geophys. J. Int.* 138, 479–494.

Sambridge, M., 1999b, Geophysical inversion using a neighborhood algorithm. II. Appraising the ensemble, *Geophys. J. Int.* 138, 727–746.

Sambridge, M., and Mosegaard, K., 2002, Monte Carlo methods in geophysical inverse problems, *Rev. Geophys.* 40(3), 1–29.

Sambridge, M. S., and Drijkoningen, G. G., 1992, Genetic algorithms in seismic waveform inversion, *Geophys. J. Int.* 109, 323–342.

Sambridge, M. S., and Gallagher, K., 1993, Earthquake hypocenter location using genetic algorithms, *Bull. Seism. Soc. Am.* 83, 1467–1491.

Scales, J. A., and Snieder, R., 1997, To Bayes or not to Bayes, *Geophysics* 62, 1045–1046.

Scales, J. A., and Tenorio, L., 2001, Prior information and uncertainty in inverse problems, *Geophysics* 66, 389–397

Scales, J. A., Smith, M. L., and Fisher, T. L., 1992, Global optimization methods for highly nonlinear inverse problems, *J. Comp. Phys.* 103, 258–268.

Schneider, W. A., and Whitman, W. W., 1990, Dipmeter analysis by a Monte Carlo technique, *Geophysics* 55, 320–326.

Sen, M. K., and Roy, I. G., 2003, Computation of differential seismograms and iteration adaptive regularization in pre-stack seismic inversion, *Geophysics* 68(6), 2026–2039.

Sen, M. K., and Stoffa, P. L., 1991, Nonlinear one-dimensional seismic waveform inversion using simulated annealing, *Geophysics* 56, 1624–1638.

Sen, M. K., and Stoffa, P. L., 1992a, Rapid sampling of model space using genetic algorithms: Examples from seismic waveform inversion, *Geophys. J. Int.* 108, 281–292.

Sen, M. K., and Stoffa, P. L., 1992b, Genetic inversion of AVO, *The Leading Edge* 11(1), 27–29.

Sen, M. K., and Stoffa, P. L., 1992c, Multilayer AVO inversion using genetic algorithms, in *Proceedings of the SEG/EAEG Summer Research Workshop*, paper 6.8, 581–589.

Sen, M. K., and Stoffa, P. L., 1994, Bayesian inference, Gibbs' sampler, and uncertainty estimation in nonlinear geophysical inversion, in *Proceedings of the European Association of Exploration Geophysicists Fifty-Sixth Meeting and Technical Exhibition*, Vienna, Austria, paper GO19.

Sen, M. K., and Stoffa, P. L., 1996, Bayesian inference, Gibbs' sampler, and uncertainty estimation in nonlinear geophysical inversion, *Geophys. Prospect.* 44(2), 313–350.

Sen, M. K., Duttagupta, A., Stoffa, P. L., Lake, L., and Pope, G., 1992, Stochastic reservoir modeling by simulated annealing and genetic algorithms: A comparative analysis, in *Proceedings of the Society of Petroleum Engineers Sixty-Seventh Annual Technical Conference and Exhibition*, Washington, DC, paper no. SPE 24754, 939–950.

Sen, M. K., Bhattacharya, B. B., and Stoffa, P. L., 1993, Nonlinear inversion of resistivity sounding data, *Geophysics* 58, 496–507.

Shalev, E., 1993, Cubic B splines: Strategies of translating a simple structure to B-spline parameterization, *Bull. Seis. Soc. Am.* 83, 1617–1627.

Shaw, R. K., and Srivastava, S., 2007, Particle swarm optimization: a new tool to invert geophysical data, *Geophysics* 72(2), V75–F83.

Singh, S. C., West, G. F., Bregman, N. D., and Chapman, C. H., 1989, Full waveform inversion of reflection data, *J. Geophys. Res.* 94, 1777–1794.

Smith, R. E., and Goldberg, D. E., 1992, Diploidy and dominance in artificial genetic search, *Complex Systems* 6, 251–286.

Spall, J. C., 1992, Multivariate stochastic approximation using a simultaneous perturbation gradient approximation, *IEEE Trans. Autom. Control* 37, 332–341.

Spall, J. C., 2000, Adaptive stochastic approximation by the simultaneous perturbation method, *IEEE Trans. Autom. Contr.* 45, 1839–1853.

Stoffa, P. L., Buhl, P., Diebold, J. B., and Wenzel, F., 1981, Direct mapping of seismic data to the domain of intercept time and ray parameter: A plane wave decomposition, *Geophysics* 46, 255–267.

Stoffa, P. L., Fokkema, J. T., de Luna Freire, R., and Kessinger, W., 1990a, Split-step Fourier migration, *Geophysics* 55, 410–421.

Stoffa, P. L., Sen, M. K., Fokkema, J. T., Kessinger, W., and Tatalovic, R., 1990b, Pre-stack shot point and common midpoint migration using the split-step Fourier algorithm, in *Proceedings of the 52nd European Association of Exploration Geophysicists Meeting and Technical Exhibition*, Copenhagen, Denmark.

Stoffa, P. L., and Sen, M. K., 1991, Nonlinear multiparameter optimization using genetic algorithms: Inversion of plane wave seismograms, *Geophysics* 56, 1794–1810.

Stoffa, P. L., and Sen, M. K., 1992, Seismic waveform inversion using global optimization, *J. Seism. Exp.* 1, 9–27.

Stoffa, P. L., Wood, W. T., Shipley, T. H., Taira, A., Suyehiro, K., Moore, G. F., Botelho, M. A. B., Tokuyama, H., and Nishiyama, E., 1992, Deep water high-resolution expanding spread and split-spread marine seismic profiles: Acquisition and velocity analysis methods, *J. Geophys. Res.* 97, 1687–1713.

Stoffa, P. L., Sen, M. K., Seifoullaev, R. K., and Pestana, R. C., 2010, Plane wave seismic data: Parallel and adaptive strategies for velocity analysis and imaging, in M. Parashar, X. Li, and S. Chandra (Eds.), *Advanced Computational Infrastructures for Parallel/Distributed Adaptive Applications*, John Wiley and Sons, New York.

Stolt, R. H., and Weglein, A. B., 1985, Migration and inversion of seismic data, *Geophysics* 50, 2458–2472.

Szu, H., and Hartley, R., 1987, Fast simulated annealing, *Phys. Lett. A.* 122, 157–162.

Tank, D. W., and Hopfield, J. J., 1986, Simple "neural" optimization networks: An A/D converter, single decision circuit, and a linear programming circuit, *IEEE Trans.* 33, 533–541.

Tarantola, A., 1987, *Inverse Problem Theory, Methods of Data Fitting and Model Parameter Estimation*, Elsevier Publishing Company, New York.

Tarantola, A., and Valette, B., 1982a, Inverse problems – quest for information, *J. Geophys.* 50, 159–170.

Tarantola, A., and Valette, B., 1982b, Generalized nonlinear inverse problems solved using the least squares criterion, *Rev. Geophys. Space Phys.* 20, 219–232.

Tikhonov, A. N., 1963, Solution of incorrectly formulated problems and the regularization method, *Soviet Math. Dokl.* 4, 1035–1038.

Tikhonov, A. N., and Arsenin, V. Y., 1977, *Solutions of Ill-Posed Problems*, W. H. Winston, London.

Upham, W., and Cary, P., 1993, Mean field theory, Hopfield neural networks and residual statics estimation, unpublished manuscript.

van Laarhoven P. J. M., and Aarts, E. H. L., 1987, *Simulated Annealing: Theory and Applications*, D. Riedel, Boston.

Van Laerhoven, Kristof, www.comp.lancs.ac.uk/~kristof/research/notes/voronoi/voronoi. pdf. Accessed Jan 1, 2013.

Varela, C. L., Stoffa, P. L., and Sen, M. K., 1994, Migration misfit and reflection tomography: Criteria for pre-stack migration velocity estimation in laterally varying media, in *Proceedings of the Sixty-Fourth Annual International Meeting of the Society of Exploration Geophysicists*, Los Angeles, CA.

Vestergaard, P. D., and Mosegaard, K., 1991, Inversion of poststack seismic data using simulated annealing, *Geophys. Prospect.* 39, 613–624.

von Mises, R., 1957, *Probability, Statistics and Truth*, English Edition, George Allen & Unwin, London.

Vose, M. D., 1993, Modeling simple genetic algorithms, in D. Whitley (Ed.), *Foundations of Genetic Algorithms II*, Morgan Kaufmann, New York, pp. 63–73.

Vose, M. D., and Liepins, G. E., 1991, Punctuated equilibria in genetic search, *Complex Systems* 5, 31–44.

Wahba G., 1990, Spline models for observational data: Soc. Ind. Appl. Math.

Walter, W. R., 1993, Source parameters of the June 29, 1992, Little Skull Mountain earthquake from complete regional waveforms at a single station, *Geophys. Res. Lett.* 20, 403–406.

Wang, L. X., and Mendel, J. M., 1992, Adaptive minimum prediction-error deconvolution and source wavelet estimation using Hopfield neural networks, *Geophysics* 57, 670–679.

Weglein, A. B., and Gray, S. H., 1983, The sensitivity of Born inversion to the choice of reference velocity – a simple example, *Geophysics* 48, 36–38.

Whitall, K. P., and Oldenburg, D. W., 1992, *Inversion of magnetotelluric data for a one-dimensional conductivity*, Geophysical Monograph Series, Society of Exploration Geophysicists, Tulsa, OK.

Whitley, D., 1989, The genetic algorithm and selection pressure: Why rank-based allocation of reproductive trials is best, in J. D. Schaffer (Ed.), *Proceedings of the Third International Conference on Genetic Algorithms*, 116–123.

Whitley, D., Starkweather, T., and Shanner, D., 1991, The traveling salesman and sequence scheduling: Quality solutions using genetic edge recombination, in L. Davis (Ed.), *Handbook of Genetic Algorithms*, Van Nostrand Reinhold, New York, pp. 350–372.

Wiggins, R. A., 1969, Monte Carlo inversion of body-wave observations, *J. Geophys. Res.* 74, 3171–3181.

Wiggins, R. A., 1978, Minimum entropy deconvolution, *Geoexploration* 16, 21–35.

Wiggins, R. A., Larner, K. L., and Wisecup, R. D., 1976, Residual statics analysis as a general linear inverse problem, *Geophysics* 41, 922–938.

Wilson, W. G., and Vasudevan, K., 1991, Application of the genetic algorithm to residual statics estimation, *Geophys. Res. Lett.* 18, 2181–2184.

Wood, W. T., 1993, Least squares inversion of field seismic data for an elastic 1D earth, PhD dissertation, The University of Texas, Austin.

Wood, W. T., Stoffa, P. L., and Shipley, T. H., 1994, Quantitative detection of methane hydrate through high-resolution seismic velocity analysis, *J. Geophys. Res.* 99, 9681–9695.

Yagle, A. E., and Levy, B. C., 1983, Application of the Schur algorithm to the inverse problem for a layered acoustic medium, *J. Acoust. Soc. Am.* 76, 301–308.

Yagle, A. E., and Levy, B. C., 1985, A layer-stripping solution of the inverse problem for a one-dimensional elastic medium, *Geophysics* 50, 425–433.

Zhao, L. S., and Helmberger, D. V., 1994, Source estimation from broadband regional seismograms, *Bull. Seism. Soc. Am.* 84, 91–104.

Ziolkowski, A., Fokkema, J. T., Koster, K., Confurius, A., and Vanboom, R., 1989, Inversion of common mid-point seismic data, in P. L. Stoffa (Ed.), *Tau-p: A Plane Wave Approach to the Analysis of Seismic Data*, Kluwer Academic Publishers, Boston, pp. 141–176.

Index

Printed in the United States
By Bookmasters